Mônica de Oliveira Pinheiro Silva

Geometry learning mediated by Cabri-Géomètre software

Mônica de Oliveira Pinheiro Silva

Geometry learning mediated by Cabri-Géomètre software

Didactic-pedagogical relations in the teaching of Geometry with the help of Cabri-Géomètre software

ScienciaScripts

Imprint

Cover image: www.ingimage.com

This book is a translation from the original published under ISBN 978-3-330-76267-1.

Publisher:
Sciencia Scripts
is a trademark of
Dodo Books Indian Ocean Ltd. and OmniScriptum S.R.L publishing group

120 High Road, East Finchley, London, N2 9ED, United Kingdom
Str. Armeneasca 28/1, office 1, Chisinau MD-2012, Republic of Moldova, Europe
Managing Directors: Ieva Konstantinova, Victoria Ursu
info@omniscriptum.com

Printed at: see last page
ISBN: 978-620-8-38654-2

To my love and partner Alexandre, for his unconditional love and support, essential fuel for the realisation of this work.

To my two princesses and daughters Julia and Luiza, for their big smiles, long kisses, vital signs for my well-being, and for the maturity they have developed throughout this journey to understand absences.

My love for the three of you is infinite.

ACKNOWLEDGEMENTS

To God who gave me life.
To my father Nilo (in memorian) for showing me by example that nothing is impossible when you do it with determination and hard work.
To my mum Stella for teaching me that things aren't easy in life and that we shouldn't give up.
To my brothers for their admiration and trust in me, and for their valuable words in times of doubt.
To my second parents Gustavo and Maria for their love and support at all times.
To my other family members for giving me the certainty that I would be able to achieve this feat.
To Prof[1] Dr[1] Neuza Bertoni Pinto for her teachings, her patience and her committed way of pointing me in the right direction.
To Prof[1] . Dr\ Zélia Milléo Pavão and Prof[1] . Dr[1] . Dilmeire Sant'Anna Ramos Vosgerau who were part of the qualifying exam board for their criticism and valuable suggestions. Especially Professor Zélia, who was my first supervisor and showed me, through her attitude, that rigour and a critical vision are basic requirements for the formation of a researcher. To Professor Marilei de Cássia Vicentini Amaro da Luz for her joie de vivre and love of teaching, which have served as an inspiration and example to me.
To Professors Marco Aurélio Kalinke, Rosana Romanó and Silvana Zilli for contributing directly or indirectly with their teachings to my research.
To the directors Guido Armando Straube and Soraya Yuri Sunaga for their trust and for allowing this research to be carried out.
To the students for their joyful participation in this work.

SUMMARY

The object of this study is the didactic-pedagogical relationships that occur in maths classes in the teaching and learning of geometry in the sixth grade of primary school, mediated by the Cabri-Géomètre software. The research aimed to understand how didactic-pedagogical relationships are established between teacher, student and mathematical knowledge, in the context of the geometry teaching and learning process, using the Cabri-Géomètre software. The analysis sought theoretical support from Guy Brousseau (1986), who underpins the theory of didactic situations, especially the didactic contract. In addition to this author, this study was theoretically based on Fainguelernt (1999), Zuin (2001) and Pavanello (1989). Valente (1999), Papert (1994), Sancho (2006), Kenski (2007) and Gravina (1989) were also consulted. The study was based on qualitative research, prioritising the context of the relationships and interactions that took place in a geometry learning environment, supported by dynamic geometry software, involving 37 sixth grade students from a public school in Curitiba. Six observation sessions were carried out on maths lessons that dealt with the teaching of quadrilaterals. Analysing the data showed that the students were familiar with using computers and did so skilfully, but had difficulties dealing with Cabri-Géomètre. This suggests a "student-machine" paradox, i.e. even though they have the skills to deal with the machine, the student is challenged by the skills provided by the Cabri-Géomètre software. Faced with this situation, new teaching contracts emerge to replace the previous ones. In general terms, the study shows that the didactic contracts established between student, teacher and knowledge, when broken, open up as an opportunity for resumption and new interactions between teacher and student in search of meaningful learning, which requires an open and flexible attitude on the part of the teacher.

Keywords: Geometry. Primary school. Cabri-Géomètre. Teaching contract.

SUMMARY

CHAPTER 1

INTRODUCTION

My personal story is certainly intertwined with my professional life. I remember as a little girl, aged six, chasing my mum around the house with my older sister's book in my hands, asking her about the letters, their sounds and meanings. I also remember that at the age of nine I would steal chairs from the kitchen in my house, line them up in the backyard, put a doll on each chair and "give lessons" to the dolls. And the lessons were in maths, with multiplication tables. What fun that was! Since then, I've never stopped doing maths.

When I was in primary and secondary school, I took part in study groups, and my job was to teach maths to my classmates. At the time, I was already intrigued by the fact that so many of my classmates didn't understand a subject that was so logical and easy for me to grasp. I was absolutely certain that I was going to be a maths teacher, because I knew that if I could teach my classmates, then I could study to be able to teach maths to more students.

At the end of the 1980s, while still a university student, I took part in projects promoted by the Federal University of Santa Catarina, one of which consisted of teaching maths to the civil servants who worked there. I was surprised at how little they knew about basic maths, from the four operations to the difference between a metre and a square metre.

My professional history is that of a maths teacher who has worked in primary and secondary schools, in both public and private schools. As a professional, I experienced changes in the Brazilian education system that had didactic-pedagogical implications for my teaching practices. One of them was the need to take the students' reality as a starting point. Regardless of whether the approach was conservative or innovative, there was a lack of understanding of what was being said in the classroom, and a clear perception of the disconnection between the maths being taught in the classroom and their reality.

When developing activities with the students, I saw two different scenarios: the real classroom situation and the theory proposed in maths textbooks. I felt a significant gap between theory and practice (a topic much discussed during my Maths degree course). The reference that came as guidance to the teacher, in the textbooks, assumed a student prepared for the grade, naturally motivated to learn the content. However, in the context of the classroom, my experience showed another type of student: unprepared, without the necessary foundation to face the grade. Most of the students needed a stimulus, a motivation to learn, and wanted to know why they were learning this or that content. They were curious about its applicability in everyday life.

The restlessness that accompanied me manifested itself even more in this conflict, this duality: theory and practice. I couldn't help wondering what I could do to bridge this gap. More specifically, in the subjects I worked on, the approach given to geometry made me feel uneasy. The students didn't have the skills to work with it, either because of the concepts and nomenclature they didn't know, or because they had little (or no) ability to construct basic elements of geometry, let alone their properties.

These concerns about teaching and learning maths are not new, as I have been sharing them with other teachers of the same subject, whether they are in primary, secondary or higher education, who are also experiencing this anguish.

As well as looking for ways to give meaning to the maths I was teaching, when I worked with geometry, I felt the problems were greater. Why? Because I had geometry in the syllabus and in the annual planning, but when it came time to teach it, to get my students "involved" in this area, the year was already over.

Throughout my career as a teacher, I have had to live with a rather unfavourable reality in relation to Geometry. Maths teaching in primary and secondary schools has shown deficiencies in this area. Several reasons can be given for this problem, such as the Euclidean approach adopted by textbooks, which are very theoretical, and the small number of lessons dedicated to teaching Geometry, which is usually taught at the end of the school year, as I have experienced many times.

This concern about the "unbinding" of geometry from the curriculum guided my work as a teacher. I came to believe that geometry would lose its place among the contents once and for all.

With the arrival of computers in all sectors, including the school world, new possibilities arose. Accompanying the change that computers could offer in terms of learning, of restructuring lessons under this new approach, I came to believe in new perspectives for the treatment of geometry.

My hopes of reducing the difficulties of teaching and learning geometry began to take shape when I came into contact with software, especially dynamic geometry software called Cabri-Géomètre.

Cabri-Géomètre was created at the Joseph Fourier Institute in Grenoble, France. The acronym Cabri comes from the French, Cahier de Brouillon Informatique, which means "Computerised Sketchbook". A group of experts worked for four years to develop this software, under the coordination of Jean Marie Laborde and Frank Belleimain (BONGIOVANNI, 1997).

The Cabri-Géomètre software consists of an interactive drawing book that offers alternatives for teaching geometry, allowing experimentation, observation and analysis of constructed geometric figures. With it we can draw geometric figures, measure segments and angles, determine geometric locations, among many other functions.

By introducing computers into the classroom, many horizons open up for an improvement in teaching and learning processes. However, any process of change will depend on the teacher's attitude towards this challenge. An educator who reflects on his or her practice realises the changes that occur with students as a result of social, political and cultural relations over the years. And this teacher, realising the changes, will have to adapt his or her practice, didactics and ways of approaching topics in the classroom, with the aim of awakening in students an investigative, critical and autonomous attitude, in line with the current needs of an individual able to face the challenges that society imposes on us.

It was with this in mind that I began to incorporate the use of Cabri-Géomètre software into my geometry lessons. I realised that the students were motivated by the ease with which they could draw, erase and redo geometric objects, but there were also gaps in their basic knowledge of geometry. I had to restructure my lessons in order to cover the current themes (of the series) with the software, while previously giving the students a grounding in the basic concepts of geometry.

It was with this motivation regarding the didactic experience with the software that I entered the Master's programme in Education. In the courses I took, the focus was, as far as possible, on my research object, which I imagined to be the Cabri-Geomètre software. When I focussed my reading on research already carried out on the use of this device, I found that various studies such as those by Fonseca (2001), Bertoluci (2003) and Zulatto (2002), among others, had concluded that the software was an effective tool for teaching geometry in terms of motivation, agility and learning.

However, I was faced with an impasse: my research would have to look for something else or take a different approach to that presented by the research I had consulted. Based on this principle, when I studied trends in Maths Education a little more, I came across the Theory of Didactic Situations, developed by French maths educator Guy Brousseau. In this theory, Brousseau identifies and details the relationships and interactions that occur between teacher, student and knowledge in a teaching and learning situation, which completed the basis that was missing for my study.

1.1 PROBLEM

Guy Brousseau's theory of didactic situations helps us to understand the relationships between students, teachers and knowledge in a teaching and learning situation. This research encompasses the above issues and incorporates the computer into this context.

Fainguelernt (1999) warns of the neglect with which geometry has been treated and points to the use of computers as a viable alternative to bridge this gap. The author emphasises the formative role that geometry plays in the construction of knowledge, focusing on the need to take the teaching of geometry more seriously. Fainguelernt states that,

> through different strategies used in the teaching-learning process of geometry, the learner has the possibility of developing the ability to activate their mental structures, facilitating the transition from the stage of concrete operations to the stage of formal operations (FAINGUELERNT, 1999, p. 22).

The reality of "neglect" in relation to the teaching of geometry is spreading and its consequences are quite damaging, having repercussions on teacher training and, consequently, on their performance in teaching in the early grades. This issue is the subject of master's research carried out by Nunes da Silva (2006), who interviewed first to fourth grade teachers and investigated what is taught and how geometry content is taught in the early grades. The difficulty that the teachers pointed out is their training, which does not cover maths in the minimum depth necessary to be able to work and develop activities that give students contact with geometry. Among other difficulties, one teacher, whose acronym represents her name is SU, says in her testimony:

> There is the geometry of the students' daily lives. We teach this when it's related to other content or facts studied. But it is left until the end of the year because of the students' greatest need, which is to learn how to do operations. (Oral testimony, in SILVA, 2006, p. 94)

This statement portrays a little of the reality of how geometry is treated in the early grades. We can see that this student, who is not allowed to learn about geometry at this level of education, will miss out on much of the opportunity to understand, describe and interrelate with the space in which they live. This gap, which begins in the early grades, is unlikely to be filled, even more so if we consider that in the following grades, geometry is often still not taught in the way it should be.

If teachers are unaware of the theoretical assumptions surrounding geometry, they won't be able to give it its due value, nor will they be able to consider it necessary enough to treat it with the same importance as arithmetic or algebra. Silva (2006, p.124) emphasises the importance of the teacher first getting to know a little more about the subject of geometry, and from there maturing to the importance of bringing it into the classroom, he says: "Geometry is the axis of content that has the role of preparing the student for the way of thinking, of exercising rationality and not just transmitting geometric knowledge".

Silva's (2006) work confirms what is experienced, i.e. realising that the gap involving geometry brought to school is opened up very early on and increases throughout the following grades.

The need to re-signify the teaching of geometry has been pointed out in research for decades. Pavanello (1989) has already pointed out some of the factors that have contributed to the abandonment of geometry. The author defends the teaching of geometry and sees it as a progressive process. According to the author,

> geometry presents itself as a fruitful field for the development of the "ability to abstract, generalise, transcend what is immediately sensible" [...] offering conditions for successive levels of abstraction to be reached (PAVANELLO, 1989, p. 182).

Thinking that geometry can be brought back into the classroom, it seems feasible to want to bring it back with a dynamic and interactive approach. One resource that makes this possible is the computer. We believe that the machine for the machine's sake will not bring anything significant in terms of improving the teaching and learning process, nor the geometry content. However, if the teacher's relationship with this resource is based on a vision of adding possibilities, then its use is justified.

Valente (1999) elucidates the difference between using the computer while maintaining the same traditional procedures, thus following the instructionist paradigm, which uses the computer as a resource that favours the traditional method, without interaction. According to Valente, "When the computer transmits information to the student, the computer takes on the role of a teaching machine and the pedagogical approach is instruction aided by it" (VALENTE, 1999, p. 2). On the other hand, if the computer is used as an aid to the construction of knowledge, its use is justified. For the author,

> When the student uses the computer to construct his or her knowledge, the computer becomes a machine to be taught, providing conditions for the student to describe the resolution of problems, [...] reflect on the results obtained and refine his or her ideas by searching for new content and new strategies. (VALENTE, 1999, p. 3).

This statement reinforces the view that it is in the hands of the teacher how the computer will be incorporated into their practice. It is impossible to deny the arrival of computers and their growing use in all sectors. Faced with this trend, we believe it is necessary for teachers to recognise this evidence and consider the possibility of incorporating it into their practice. Making computers available to students does not mean that this will be enough to guarantee meaningful learning, or that this scenario will lead to constructivist practice.

Valente (2005) states that the technical facilities offered by computers make it possible to explore countless pedagogical activities. On the other hand, he questions whether these activities really contribute to the process of constructing knowledge. The author draws attention to the subtle difference between carrying out activities with the computer and carrying out activities that may in fact require some reflection and contribute to the apprehension of concepts in a meaningful way. In relation to this subtlety, the author points out:

> The student may be searching for information on the Internet, in the form of text, video or graphics, pasting it into a multimedia creation, but without having criticised or reflected on the different contents used (VALENTE, 2005, p. 23).

The author emphasises that teachers have the potential to combine technical resources with pedagogical knowledge for truly meaningful learning and quality education. To this end, he suggests emphasising and taking care in teacher training so that they can face up to the great challenge of incorporating computers into their practice in a meaningful way. For the author, in relation to teacher training: "It should be thought of in the form of a growing spiral of learning, allowing the educator to simultaneously acquire technical and pedagogical skills and competences" (VALENTE, 2005, p. 30).

Kenski (2007) places us in relation to technological advances and their implications for the school world. She also clarifies that technologies are not just machines connected to equipment, and considers language to be a specific type of technology. The definition of technology according to the author is:

> The set of scientific knowledge and principles that apply to the planning, construction and use of equipment in a particular type of activity is called technology (KENSKI, 2007, p. 24).

This view helps us when dealing with technological resources, avoiding the idea that if the school doesn't have the Internet or educational software, it won't be possible to develop a pedagogically innovative activity. According to the author, "newspapers, magazines, radio, cinema, video etc. are popular media with enormous social penetration" (KENSKI, 2007, p. 27).

According to Gravina and Santarosa (1998), students arrive at university without having reached the mental levels of deduction and rigour. Deductive reasoning, methods and generalisations, characteristic and fundamental processes of Geometry, are poorly mastered by students, who even have little understanding of the objects. The author proposes that once teachers accept the computer as a tool and incorporate it into their practice, care must be taken to ensure that the use of this technology is in line with a constructivist approach. To this end, the author proposes two main issues related to a practice that uses the constructivist approach:

> 1ª : Regarding the mathematical aspect: how to design activities that make students appropriate deep and meaningful mathematical ideas [...]
> 2a: With regard to the cognitive aspect: how can we ensure that these activities place students in attitudes in tune with the processes that are natural to the cognitive development of the subject? (GRAVINA; SANTAROSA, 1998, p. 5).

Faced with the questions proposed by Gravina and Santarosa (1998), the teacher who sets out to work in a meaningful way, using the computer as a tool, must be prepared to reflect on their "new" practice, focusing on these questions. The author goes beyond the issue of using computers and proposes learning mathematics in computerised environments, pointing out the characteristics that make up a constructivist computerised environment. These are: a dynamic environment, an interactive environment and an environment for modelling or simulation. In this context of learning in a computerised environment, some software is proposed, including a geometry software called Cabri-Geomètre.

The Cabri-Géomètre software allows students to get in touch with the language of geometry, because "the construction menu interface is in the classical language of geometry. Drawings of geometric objects are made from the properties that define them" (GRAVINA; SANTAROSA, 1998, p. 14).

From what has been said so far, a reality emerges where there is an empty space in the treatment given to geometry, since there is a strong and irreversible trend towards the use of computers in classrooms, and the focus of this study will be on the use of that tool, in order to contemplate the constructivist approach, seeing the opportunity to involve these components for the development of the research.

It is therefore within this space of needing to analyse the teaching of geometry, supported by geometry software, that we formulated the following question:

In the context of the process of teaching and learning geometry with the Cabri-Géomètre resource, how are the didactic-pedagogical relationships between teacher, student and mathematical knowledge established?

1.2 OBJECTIVES

1.2.1 General objective

To analyse the didactic-pedagogical relationships that occur between teacher, student and knowledge in the teaching of geometry mediated by Cabri-Géomètre.

1.2.2 Specific objectives

a) describe the forms of mediation that took place during the activities with Cabri-Géomètre;
b) identify the roles of the teacher and students when using the software;
c) to characterise and analyse the didactic contract between teacher, student and knowledge during the use of the software;
d) to describe the process of teaching and learning geometry through the use of the Cabri-Géomètre software, in the light of the Theory of Didactic Situations.

1.3 METHODOLOGY

In line with the proposal to study the interactions that take place in a learning environment mediated by the use of software, some steps had to be taken to achieve the objectives presented. As we were already teachers at the school and using the software, we asked the teacher of the other class if she would agree to contribute to this research. She agreed and we then spoke to the school headmaster, explaining what the research would be like, the subjects involved and the environment in which it would take place. The headmaster allowed the research to take place at the school and the teacher, according to her planning, told us which days of the week she would be teaching geometry to the selected class.

The research, which took a qualitative approach, was carried out with students in the 6th grade[a] of a public school in Curitiba that uses its own teaching materials, which are produced by the institution's Centre of Excellence. This institution has approximately 2,150 students enrolled at all levels of education, from nursery school to university.

Data collection consisted of recordings of classes held in the Computer Laboratory of the research institution. At each lesson, the class teacher put a tape recorder in the pocket of her coat to record what happened. In addition to this recording, we made notes of what happened in each lesson in a "logbook", reporting our perceptions and later making a more reflective record of the observations made.

In this empirical work, the records were centred on key elements of the observations: the student, the teacher, the learning environment (computer lab), 6th grade geometry content and the use of Cabri-Géomètre software.

1.4 WORK STRUCTURE

Firstly, the Theory of Didactic Situations is addressed, as well as an analysis of its structuring concepts; in addition, an approach to the different types of didactic situations proposed by Brousseau and highlighting the didactic contract as a fundamental element for understanding the interactions that occur between teacher, student and knowledge in a didactic relationship.

Next, we look at geometry in primary school. We briefly describe the trajectory of geometry in Brazilian schools, how it is currently conceived and treated in the school context, especially in the final grades of primary school. Drawing on research already completed on the teaching of geometry in Brazil, as well as on the National Curriculum Parameters (PCNs) for Primary Education, in this part we try to find out about the questions and conclusions about the teaching of geometry in Brazil, consulting theses and dissertations as well as research on geometry debated in the GT19 of Anped .[1]

This is followed by a discussion of technological advances and the implications of using computers in school practices. Cabri-Géomètre, one of the components of this object of study, is then

[1] ANPED aims to develop and consolidate postgraduate teaching and research in the field of education in Brazil. Over the years, it has established itself in the country and beyond as an important forum for debating scientific and political issues in the area, and has become a reference for monitoring Brazilian production in the field of education. As for WG 19, it is as follows: Maths Education is an important space for debating selected research in this area of knowledge.

presented. Finally, the role of the teacher in using a technological tool is highlighted.

Next, the methodological elements that guided this study are detailed, characterising the approach chosen, the universe in which the data was collected, the subjects involved and the procedures used to collect and analyse the data.

Next, the data collected in the geometry lesson observation sessions is analysed and discussed. The episodes selected from the dialogues and interactions that took place during the observation sessions are discussed in this section, in the light of Didactic Situation Theory.

The Final Considerations draw conclusions from the analyses. Also added to the analyses are what was possible to observe in terms of future possibilities, such as directions and new research proposals.

Overall, this research is a study that addresses issues that are relevant to the school world. In schools where the use of computers is systematised, maths teachers will be able to share and understand even more the importance of using computers efficiently and with a focus on learning. For real progress to be made in mathematical knowledge, it is important that teachers design their activities with care to allow exploration, conjecture and recovery of basic geometric concepts.

Even in schools where the use of computers is not so efficient, this work points to the tendency to use this tool. Even if the teacher has no way of using it, the relationships implicit in the didactic contract are present, with or without the use of computers. In the moments of interaction between teacher, student and knowledge, the implicit and/or explicit relationships, the implicit rights and duties of the students and the teacher in relation to the object of mathematical knowledge are presented in this research. In this way, this study shows topics from the Theory of Didactic Situations that can lead teachers to reflect on their practice from the point of view of the didactic contract.

Finally, we believe that this research is also justified by its social relevance, in the sense that it can contribute to a greater understanding of the problems surrounding geometry, and thus awaken in educators the need to rethink maths teaching and look for alternatives to remedy this deficiency. As well as providing a reflection on the teaching of geometry, this work proposes alternatives for using the computer as a tool, and not as the "main actor" in computerised environments.

We hope that this study can contribute to improving the quality of geometry teaching, serving as a parameter and an alternative for teachers working at this level of education who feel the need to redirect their practice.

CHAPTER 2

THE THEORY OF DIDACTIC SITUATIONS

This chapter discusses the theory of didactic situations developed by Guy Brousseau. Initially, a brief biography of the author is presented, followed by a summary of the Theory of Didactic Situations, analysing its main structuring concepts. Among the core concepts of this theory, we highlight the concept of the didactic contract and its implications for the relationships that occur between teacher, student and knowledge, taking into account the importance of understanding these interactions present in the teaching and learning processes of Geometry, mediated by the use of the Cabri-Géomètre II software, the object of this study.

2.1 MATHS EDUCATOR GUY BROSSEAU

French educator Guy Brousseau began his career in 1953 as a primary school teacher. At the end of the 1960s, after graduating in Maths, he joined the University of Burdeos as a lecturer. In 1986, he completed his doctorate and, in 1991, he became a professor at the new University Institute for Teacher Training (IUFM) in Burdeos, where he worked until 1998.

At the beginning of the 1970s, Guy Brousseau emerged as one of the most original leaders and researchers in the new field of French Maths Didactics. His most notable theoretical contribution was the development of the Theory of Teaching Situations, centred on the social interactions that take place in the classroom between students, teachers and knowledge. Disseminated worldwide, his theory has served as a reference for current and more in-depth studies on the relationship between students, teachers and mathematical knowledge. More than just an excellent researcher, Guy Brousseau is also an academic who has dedicated his life to Maths Education, not only in France, but also in other countries. He has supervised more than 50 doctoral theses, contributing to the development of Maths Didactics. Guy Brousseau is currently a researcher at the IREM (Institute for Research into Mathematics Teaching) in Bourdeos, as well as Director of COREM (Centre for Observation and Research into Mathematics Teaching); he is also Professor Emeritus at the IUFM in Aquitania (Bordeaux) and Doctor Honoris Causa of the University of Montreal.

2.2 THE THEORY OF DIDACTIC SITUATIONS

In the 1970s, a group of researchers (mostly French mathematicians and psychologists) joined forces to understand and interpret the phenomena and processes linked to Maths Education, a field of study known in France as Maths Didactics. Convinced of the need for a specific theoretical body of mathematical knowledge, they began in-depth investigations and developed substantial theories for the field of maths teaching and learning, among them[2] , the "Theory of Didactic Situations", developed by Guy Brousseau, which seeks the genesis and understanding of the phenomena that occur in the teaching and learning of school maths. Based on a constructivist conception (in the Piagetian sense), in which the student learns by being challenged by a situation, producing new responses to an environment full of contradictions and difficulties, Brousseau (1996) calls a "didactic situation" a situation constructed with the intention of enabling the student to learn a particular mathematical content.

For Brousseau (1996), a didactic situation is a set of relationships explicitly and/or implicitly established between a student or a group of students, instruments or materials and the teacher, with a view to learning some knowledge. For the author, students only construct knowledge when they get involved with the situation, i.e. when they get personally involved with the problem proposed by the situation, and there is therefore a return on the part of the student. This means that the student enters the "game", i.e. enters the "mathematical functioning".

It is, therefore, a planned situation with the intention of enabling the student to construct a particular piece of knowledge. According to Brousseau (1996), some of these situations require prior knowledge and the necessary schemes. Others offer students the chance to construct new knowledge on their own. According to D'Amore (2007):

> This type of theory, let's not forget, has maths as its reference and, therefore, when we talk about knowledge, we always imply that we are talking about mathematical knowledge; now, mathematical knowledge, as its own characteristic, includes concepts, but also systems of symbolic representation, processes of development and validation of new ideas (D'AMORE, 2007, p. 237).

The Theory of Didactic Situations involves different types:

a) action situations;
b) formulation situations;
c) validation situations;
d) institutionalisation situations.

An action situation is the type of situation in which students use their most immediate

---

[2] Some of the theories that also stand out among those developed by the French Maths Didactics Group are: Theory of Didactic Transposition, developed by Yves Chevallard; Theory of Conceptual Fields, developed by G. Vergnaud.

knowledge and procedures to solve a problem. In such situations, the experiential and argumentative aspects prevail, while the theoretical aspect of the concepts involved remains in the background. This is the case where the student is able to provide the solution to a proposed problem, but is unable to explain and/or explain the mechanisms and elements used in its development.

A formulation situation is when the student starts to use some kind of theoretical elaboration to solve the problem, presenting more elaborate reasoning than an experimental mechanism, and for this it is necessary to use previous information. In the formulation situation, the knowledge being developed does not require justification and control of actions. The aim of this stage is to exchange information, consisting of moments in which the student explains the tools used and the solution found. "This is the case where the student makes statements without the intention of judging the validity of the knowledge, although they implicitly contain validation intentions" (PAIS, 2002, p. 72). In order for students to make progress in solving a problem, they need to deepen their reflective stance, seeking justification for the validity of the statements made, even if only in an internalised way.

The validation situation is more focused on rational argumentation, where the question of the veracity of knowledge is now important. In a validation situation, the student is able to use some explicit theoretical models and schemes to solve the proposed problem, and it is possible to present theoretical arguments in a much more elaborate way, such as proof mechanisms, and knowledge is used for this purpose. In this situation, the main characteristic is the debate about the veracity of assertions. As a result, interactions with the environment are organised. In this endeavour and intellectual motivations, together with the procedures of argumentation about knowledge, the student can contest or even reject propositions that they don't yet understand, and can experience the condition of doubt, going in search of theoretical contributions that make them agree with and accept a proposition, or debunk it with specific and true arguments. This requires verifications and therefore explanations of the theory used, as well as explaining the means used in the demonstrative processes. The student's intellectual work is not just about information about knowledge, but also involves affirmations, elaborations and statements about the validity of knowledge (PAIS, 2002). In this way, theory works and shows itself in scientific debates or discussions between students as a means of establishing proof or rejecting it.

In the situation of institutionalisation, the role of the teacher comes to the fore, with the purpose of bringing together the debates, arguments and ideas that have arisen, making it official and conferring a status on the knowledge. It is in this situation that knowledge moves from the individual and private level to the historical and cultural dimension of scientific knowledge. These situations aim to establish and give an official character to the knowledge that emerges during class activity.

When it comes to the transition from individual knowledge to social, constructed and validated knowledge, it's worth highlighting the differentiation that exists in the social dimension of subjective knowledge, since this knowledge is now (in institutionalisation) accepted by the environment with the status of non-localised knowledge. Once constructed and validated, the new knowledge will only become part of the class heritage once the teacher has explicitly established and agreed on the cognitive status of the knowledge. Here we emphasise the need for rigour on the part of the teacher when institutionalising, so as not to do it at the wrong time. If done too early, institutionalisation interrupts the construction of meaning and can prevent proper learning, creating difficulties for the students and also for the teacher. If done too late, it opens the door to distorted and erroneous interpretations, making learning more difficult.

According to Pais (2002, p.74), "Institutionalisation only makes sense when the student understands the meaning of the content and realises the need to integrate their knowledge into a broader theory". In this case, the institutionalisation of knowledge comes into play, for example, when checking that problems have been solved, when taking stock of the activities carried out in class, either individually or in groups, and it is up to the teacher to select the formal aspects of the content, which come to be seen as knowledge that is culturally accessible to the student. It is in the institutionalisation situation that the teacher draws conclusions from what has been produced by the student, recapitulating, systematising and ordering what has been produced, with the aim of establishing relationships between what has been produced by the student and their culture. According to Brousseau:

> The role of the teacher is to institutionalise! Institutionalisation takes place both in a situation of action - the value of a procedure is recognised and it becomes a reference resource - and also in a situation of formulation. There are formulations that will be preserved ("this is how it is said", "that should be remembered"). It's the same with proofs: you have to identify or you will be retained from the properties and objectives that you find (BROUSSEAU, 1996, p.57).

Panizza (2006, p. 39) summarises the different didactic situations, saying: "in action situations, actions are validated; in formulation situations, messages are validated; in validation situations, statements are validated". In addition, in institutionalisation situations, the official meanings and conventions of knowledge are validated.

On the phenomenon of institutionalisation, Brousseau states:

> Traditional teaching situations are institutionalised situations, but without the teacher being involved in creating meaning: you say what you want the child to know, explain it to them and check that they have learnt it. At first, researchers were a little surprised by a-didactic situations, because that was what traditional teaching lacked the most (BROUSSEAU, 1999, apud PANIZZA, 2006, p. 40).

2.2.1 A-didactic situation

It's important to note that when planning a didactic situation, according to the creator of the theory, there are moments in which the student is alone when faced with the problem to be solved, without the intervention of the teacher. Brousseau calls this moment an a-didactic situation or phase, a stage in which the student must relate to a problem based on their own knowledge, challenged by the problem "and not to fulfil a wish of the teacher, and without the teacher intervening directly to help them find a solution" (PANIZZA, 2006, p. 37).

One of the aims of maths education is to help students develop a certain intellectual autonomy and to ensure that the school knowledge they learn enables them to understand and critically situate themselves in the world in which they live. To this end, in addition to the situations proposed, designed and planned by the teacher to achieve these objectives, called didactic situations, there are many situations in everyday school life that are not under the pedagogical control of the teacher, in other words, they are moments in the learning process in which the student works independently, not suffering any kind of direct control from the teacher. These are called a-didactic situations, which are also implicit in the didactic contract. A-didactic situations occur when moments in the learning process take place without the direct interference of the teacher (in which the teacher is apparently absent), when, at a certain stage of the didactic contract, the student individually or collectively seeks to construct a new concept. In Brousseau's words:

> When the student becomes capable of putting into operation and using for himself the knowledge he is building, in a situation not foreseen in any teaching context and also in the absence of any teacher, then what can be called an a-didactic situation is occurring (BROUSSEAU, 1986, apud PAIS, 2002, p. 68).

According to Pais (2002, p.69), "the pedagogical intention characterises all the stages of the didactic process, since all the teacher's work is previously determined by objectives, methods and conceptual notions." Thus, a-didactic situations are moments of great potential, since they are geared towards trying to break away from the old practices of repetition and modeling, which still exist in traditional teaching practices. This does not mean that didactic and a-didactic situations are contradictory, but rather that they coexist in a didactic phenomenon in harmony, without one altering the other, but rather that one situation can complement the other. We emphasise that any pedagogical activity, when planned, must include moments to direct the student towards a-didactic situations. This means that students can carry out mathematical investigations, research and/or delve deeper into a particular topic, regardless of the educational system or the teacher's pedagogical intentions, and they will still be experiencing a-didactic situations.

We can't confuse an a-didactic situation with a non-didactic situation. Non-didactic situations are those that have not been planned with the aim of promoting learning. In this type of situation, the problem arises occasionally, without the intentionality of knowledge, and the teacher and student do not have a specific and typical relationship as the knowledge at stake.

Being aware of the elements contained in some didactic contracts and also of the need to evaluate, renegotiate and understand classroom phenomena, the teacher must be able to perceive, manage and enforce their role and that of the student in each contract. And within this panorama, in which indignation arises at the possibility that maths is being under-learned and misinterpreted, we formulate the following question: "How can we develop situations that encourage reflection, investigation and autonomy?" To promote a didactic situation, Brousseau (1996) explains:

> The modern conception of teaching therefore asks the teacher to provoke the desired adaptations in the student through a judicious choice of "problems". These problems, chosen in such a way that the student can accept them, must lead him to act, speak, reflect and evolve on his own (BROUSSEAU, 1996, p. 49).

The rigour required in drawing up the questions and problems to be proposed avoids interference, distortions and ambiguities in a teaching situation. One consequence of not choosing wisely when promoting appropriate learning situations is called learning by adaptation. Brousseau (1996), making an approximation with the so-called assimilation and accommodation schemes described by Piaget, defined this situation (learning by adaptation) in a context where "the student is challenged to adapt their previous knowledge to the conditions for solving a new problem. In this case, learning is expressed through the component of creativity" (Pais, 2002, p.69). The student is faced with the need to adapt their knowledge to a particular problem that has been proposed to them within a didactic situation. In adaptive learning, students are forced to adapt their previously acquired knowledge to the conditions of a new problem. This type of learning brings up another clause in the didactic contract, in which it is up to the student to learn, since the teacher is fulfilling the contract by teaching. It is also up to the student to fulfil their part of the contract, which is to learn, even if this requires adaptations and distortions. At the moment of learning, it is not up to the student to change the focus of study, but rather to construct the concept gradually and procedurally.

As for the role of the teacher, he or she is allowed to use adaptive teaching strategies and, as already mentioned, to use criteria when choosing the approach to the object of knowledge. It is also up to the teacher to search (no less carefully) for problems that allow for more than one solution, that value creativity and allow for personal strategies. In this scenario, the students' actions and reactions manifest themselves, signalling to the teacher the need (or not) for a retake, a renegotiation, and the consequent breaking of the didactic contract in force.

This renegotiation takes place when the teacher realises that something was not planned. If a change of course is necessary, improvisation consists of readjusting the plan at the exact moment when the teaching situation occurs, when the proposed objectives are not being achieved, or when something is causing the focus of the lesson to be diverted. The concern here is different, since new (unforeseen) variables have appeared. In this case, it's up to the teacher to show their ability to manage this moment, guiding their students through the new questions.

There are situations in which students try not to "admit" that something has got out of hand. When they try to understand the reason for their loss of control, they are faced with "giving back", in other words, the student feels responsible for the activity and also responsible for the result to be obtained. The stage of devolution is what is expected in a didactic situation, although it doesn't always happen. Students need to accept the idea that they are responsible for the activity, and that they should go in search of solutions to the proposed problem situation. Brousseau (1996) warns about this moment:

> For a child to understand a situation as a necessity, regardless of the teacher's wishes, requires an intentional cognitive epistemological construction. Solving the problem then becomes the responsibility of the pupil, who must endeavour to achieve a certain result (BROUSSEAU, 1996, p. 50).

There is nothing more appropriate than for the teacher, faced with this student nonconformity, to allow the student to mature and build their knowledge in an a-didactic situation. When referring to an a-didactic situation, Pinto (2003, p.102) states that "it is the moment considered the most important in learning, in which the student, working without the teacher's interference, seeks to synthesise their knowledge, develop their own methods, activate their meta-cognition".

Referring to the term a-didactic, coined by the French mathematician Brousseau, Panizza (2006, p. 37) emphasises its importance, since often "what is necessary is confused with what is possible to use as procedures to solve a problem and, consequently, the knowledge that must or must not be put into practice in order to master the situation is confused".

An example of this is when the student has to solve the situation: "Put two sets of trolleys together and find the total of the trolleys". The student will not always use the calculation of the two sets. They can put the two sets of trolleys together and count the new set formed. They can also use representations, such as toothpicks or their fingers, to count the total without calculating. By using his inventiveness, the student is experiencing an a-didactic situation.

The teacher's non-intervention in this process makes it possible for the student's learning to progress. Entering the a-didactic phase gives rise to the concept of devolution, also developed by Brousseau, i.e. "the act by which the teacher makes the student accept responsibility for a learning situation (a-didactic) or a problem, and he himself accepts the consequences of this transfer" (BROUSSEAU, 1996, p.50).

2.2.2 Returning a situation

Brousseau warns about the criteria that the teacher must have when preparing didactic situations, so as not to get lost during the game (the lesson), or so that the situations aren't so obvious that they aren't exciting. In addition, the teacher, being implicitly responsible for the learning situation, must have the ability to make the student see the task as their own, to make them feel motivated to the point of taking responsibility for the problem, and to go in search of the answer, as a challenge of their own.

> The teacher's job, then, is to offer the pupil a learning situation so that they can develop their knowledge as a personal response to a question, and make it work or modify it in response to the demands of the environment and not to the wishes of the teacher (BROUSSEAU, 1996, p. 49).

When the student accepts the proposed situation and takes personal responsibility for solving the problem presented to them in the didactic situation, the student is said to have achieved the return of the situation. It is not enough for the teacher to simply "communicate" a problem to a student so that it immediately becomes their problem, even if the student feels solely responsible for solving it. Nor is it enough for the student to accept responsibility for solving the problem. There must be responsibility on the student's part, submitting to the consequences of solving a given question. For Brousseau (1996), devolution is the activity through which the teacher manages to achieve both results, i.e. that the problem proposed by the teacher is accepted by the student, and once accepted, that the student feels responsible for the issue, assuming the consequences of this transfer.

We can also look to D'Amore (2007, p. 81) for another contribution to the understanding of the return of a situation: "The return is therefore a situation in which the student 'works' in a scientific way, and not just in response to stimuli external to the situation, of a didactic type, for example". Giving back is important because it drives the learning process, getting the student into a context where they want to think, to function in a mathematical way in the face of a problem they want to solve. On the one hand, the student knows that being in the classroom, with the teacher formulating a problem, there is an intention to learn, but in order to do so, they will have to face the problem on their own initiative, their own curiosity, without the need for the teacher's intervention, or without some didactic reason. In order for learning to take place, one way we believe to be assertive is to ask students good questions, giving them opportunities for good answers.

As Brousseau (1986, p. 53) states, "the teacher has to be concerned not with the communication of knowledge, but with the return of the appropriate problem. If this return occurs, the student enters the game and, if he ends up winning, learning has taken place".

The process that complements feedback is therefore the institutionalisation of knowledge.

After the appropriate conjectures, debates, doubts and theorising, the time has come to articulate the knowledge that students put into play when solving problems, in order to establish and give official status to this knowledge.

2.2.3 Teaching obstacles

The idea of didactic obstacle was introduced by Brousseau to further flesh out the themes researched and debated in the field of Maths Didactics, inspired by the idea of epistemological obstacle that was first described by the French philosopher Gaston Bachelard, presented in 1938 in his work "The Formation of the Scientific Spirit".

The term obstacle already suggests the condition of making something difficult. According to the Aurélio dictionary (OBSTÁCULO, 1993, p.1211), the word obstacle means: "embarrassment, difficulty, impediment, hindrance, obstacle, barrier".

Epistemological obstacles are old knowledge, crystallised by time, which resist the installation of new conceptions and which threaten the intellectual stability of those who hold this knowledge.

For D'Amore (2007), epistemological obstacles are actually manifested in the very act of knowing, intimately, and that slowness and disorder appear out of a kind of functional necessity. It is precisely at these moments that causes of stagnation and regression arise, and it is there that causes of inertia, called epistemological obstacles, are identified.

According to Pais (2002), to approach the epistemological concept, with reference to the formation of mathematical concepts, obstacles appear more intensely in the phase of learning and synthesising knowledge than in its historical record. The author also states that "didactic obstacles are knowledge that is relatively stabilised on an intellectual level and which can hinder the progress of learning school knowledge" (PAIS, 2002, p.44). In other words, old knowledge acts as an opposing force to the realisation of new learning. In addition, the author recommends a certain criterion in the use of this term (didactic obstacle) so as not to make generalisations, since its use is more pertinent on a pedagogical level, to the detriment of the specific character of the historical context of the sciences.

The sources of school learning difficulties are diverse, and deserve a close look in order to identify them according to their specific characteristics and seek ways of minimising them. If we consider the subject of Mathematics, for example, with its aim of promoting the learning of concepts in lessons, it is therefore necessary to understand how intellectual and cognitive reorganisation takes place, so that the new knowledge can come into harmony with previous knowledge, and it is at this point that obstacles manifest themselves.

A first example of a didactic obstacle would be in the study of arithmetic, when it comes to learning the product of two positive numbers, which is always greater than each part. Thus, when this concept is extended to rational numbers, the concept is not valid.

Another example, still related to operations with rational numbers, is when you divide a positive integer by a rational number less than one, resulting in a number greater than the dividend. This is a didactic obstacle, as it is based on the concept that dividing always decreases.

When it comes to learning spatial geometry, the didactic obstacle manifests itself when working with a perspective representation. We can take figure 1 as a reference.

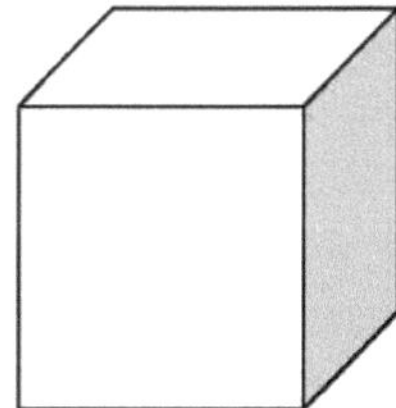

Figure 1 - Representation of a cube
Source: The author

A cube represented in parallel perspective usually appears with the upper face represented by a parallelogram and not a square, where the angles are not right angles when measured on the surface of the paper, but represent right angles, as they are the face of a cube, formed only by right angles.

If the central theme of this study is geometry, this type of barrier draws our attention in a special way, since the use of Cabri-Géomètre software can be an agent for minimising this type of didactic obstacle, since working with solids in the third dimension allows students to see all the angles of the geometric solid.

Studies carried out by Baldy (1987) apud Pais (2002) have shown that drawing can present difficulties when learning geometry, drawing our attention to didactic obstacles. Work carried out by the Geometry group at the Institute for Research into the Teaching of Mathematics (IREM) in Montpellier has also shown that there are difficulties that students can encounter in the study of spatial geometry when they need to visualise and interpret a drawing in perspective, and that there can be confusion between the properties of the traces of the drawing itself and the geometric elements represented by them, such as the example of the cube seen in perspective, mentioned above.

According to Pais (2002), there are different types of didactic obstacles, including the simplified way in which content is presented in textbooks, in which the stage at which it is intended to be formalised does not correspond to the challenges of the cognitive phenomenon, without the rigour and abstractions necessary for this stage of learning. Another type of didactic obstacle is generalisation.

Generality can become an obstacle when there is a hasty attempt to generalise an idea, a concept that is still in the elaboration phase, stuck in a pre-reflective understanding. This obstacle occurs when a concept is taken to the next level.

field of science, without the methodological criteria and care of research. The generality that can become a didactic obstacle is that which is interpreted as a type of knowledge that, in its attempt to have an overview of the whole, ends up getting lost in its superficiality. This would not be the case with that generality which is necessary at the right time, such as the generality of a theorem, which is only justified as a synthesis of the regularity existing between an infinite number of particular cases, and is not vague or superficial knowledge.

To complement this analysis of didactic obstacles, we use the words of Igliori (1999):

> a) the conceptions that cause obstacles in maths teaching are rarely spontaneous, but come from previous teaching and learning;
> b) the mechanisms that produce obstacles are also producers of new knowledge and factors of progress;
> c) the obstacle is related to a knot of resistance that is more or less strong according to the students, the teaching received, because the epistemological obstacle is often broken down into obstacles of other origins, notably the didactic one. (GLORIAN (s/d), apud IGLIORI ,1999, p. 110).

We understand that didactic obstacles occur in a learning situation, most of which is mediated by the teacher, because there is still an attachment to the idea that knowledge, the construction of concepts and learning occur in the classroom, in the presence of the teacher. Fortunately, research in maths education, whose object of study is to understand the phenomena relating to the teaching and learning of maths, shows ways in which learning can take place in a-didactic situations, in other words, it is possible to disconnect learning from the presence and interference of the teacher in a unique and exclusive way. However, we know that it is the educator who leads the process in most classroom situations. It was then, by carefully analysing classroom phenomena, that Brousseau created the term didactic effects, which will be seen below.

2.2.4 Teaching contract: its implications and the subjects involved

A fundamental concept that permeates Brousseau's theory is the didactic contract. This concept should not be confused with the pedagogical contract, nor with the concept of the social contract, which we will define below. The concept of the didactic contract was introduced by Guy Brousseau (1986), to explain relationships that occur in the classroom involving the teacher, the student and knowledge. For Brousseau, the didactic contract refers to:

> The study of the rules and conditions that condition the functioning of school

> education, whether in the context of a classroom, in the intermediate space of the school institution or in the wider dimension of the education system (BROUSSEAU, 1986 apud PAIS, 2002, p. 77).

This concept was launched in the 1970s as part of research into the Didactics of Mathematics aimed at studying school failure, which in that decade was intensively discussed and analysed by researchers, especially by French mathematics educators challenged by the low performance of students in the subject of Mathematics, a situation that is still the subject of current studies. For Brousseau:

> In a teaching situation, prepared and carried out by a teacher, the student normally has the task of solving the (mathematical) problem presented to them, but access to this task is made by interpreting the questions posed, the information provided, the obligations imposed that are constant in the teacher's way of teaching. These (specific) teacher habits expected by the students and the student behaviours expected by the teacher constitute the didactic contract (BROUSSEAU, 1980, p. 127 apud D'AMORE, 2007, p. 101).

With this definition, we realise the comprehensiveness of the concept and its implications in terms of the expectations that are manifested bilaterally between teacher and student, and it is logically up to the teacher to have mastery and understanding of this concept in order to intervene effectively in learning situations. In addition to its comprehensiveness, we emphasise the character of movement in a reality, such as the one that exists in a classroom, which is neither stable, nor static, nor established forever, but rather an evolving reality that improves and changes. Brousseau uses the notion of the didactic contract in a way that is analogous to a game, in which the teacher is involved in the system of the student's interactions with the problem he or she is posed, this game being the didactic situation.

More generally, a definition of a contract, according to the Aurélio dictionary (CONTRATO, 1993, p. 468) is: "The act or effect of contracting; an agreement between two or more people who transfer some right or subject themselves to some obligation". From this definition, we can see that in the social and pedagogical universe that is the school, its form of organisation is structured through a network of relationships involving teachers, students, parents, administrative staff and other employees. These fit into countless contracts that are established between the different "actors" in schools. Many of these relationships and contracts are not so readily apparent to students. They go to school and are subject to the rules imposed and agreed upon by the school and their parents, who leave them "out" of this agreement, leaving them "only" to fulfil their role as a student. The teacher/student relationship is a special kind of relationship, always mediated by knowledge. It is formally elaborated and systematised, aimed at achieving this knowledge, which we understand as learning. It is from this close relationship that the didactic contract will emerge, involving student, teacher and knowledge.

Pinto (2003) seeks to clarify the different types of contracts. According to the author, the social contract proposed by Rousseau, "rejects any pact of submission by one of the contracting parties, and likewise any authority that emerges from the privileges of nature or the rights of the strongest" (PINTO, 2003, p. 96). In this contract, according to the author, the will of the community prevails. The term pedagogical contract, on the other hand, according to Pinto (2003), appeared in the pedagogical vocabulary as a result of experiments called the "Dalton Plan", in which teachers used a "rationalisation" of the teaching programme, organised into minimum units of study. In this experience, students were free to carry out their tasks and the teacher only intervened to guide the students' work. The tasks were carried out on a monthly basis, with the aim of breaking away from the over-intellectualism of the classes and seeking to individualise teaching. This experiment was criticised for being too close to traditional programmes.

According to Pinto (2003), another conception of the pedagogical contract is found in Filoux (1974), apud Pinto (2003), in which there is a need for "mutual consent" between teacher and student regarding the rules that are established in a didactic relationship. In this contract, it is (apparently) impossible for the subjects involved to participate in changes to the rules that are intended to regulate the roles of teacher and student in relation to the institution. In this sense, as the author observes (p. 99), "the pedagogical contract implicitly brings power relations whose negotiations, not always made explicit, are already previously established in the institutional contract".

Seeking to clarify the differences between the didactic contract and the pedagogical contract, Pinto (2003) states that in the pedagogical contract, based more on traditional teaching, the norms are explicitly given by the teacher's knowledge (grades, punishments, rules) and it is up to the student to obey and comply, so as not to be punished. In the didactic contract, the main focus is on the nature of the knowledge involved in a teaching situation, in which teacher and student interact, each in their own role. The author also states that, according to the definition of the didactic contract, if one of its functions consists of dynamising the interactions established by the teacher and the student with knowledge, its rules cannot be unilateral, since they depend on two important "poles" of the didactic relationship established between student and teacher, in search of the construction of knowledge on the part of the student, and in the elaboration of methods, strategies for the construction of knowledge on the part of the teacher. "The democratic sense of the contract is guaranteed by situations in which the student is challenged to produce new knowledge" (PINTO, 2003, p.104).

In the classroom, the didactic contract refers to the obligations, attitudes, postures and actions of the students, which are expected by the teacher, and also those of the teacher, which are expected by the students.

The attitudes on both sides, and those expected by each of them from the teacher as expected by the students and from the students as expected by the teacher, are what make up the didactic contract.

In maths classes, the most common pedagogical practice still seems to be one in which the teacher complies with the rules of a contract specific to lectures, which in fact characterise a pedagogical contract and not a didactic contract, according to what Brousseau proposes.

This posture of the pedagogical contract begins with the teacher, who, when introducing a new concept, does so without showing its applicability, nor without referring to the new subject in a contextualised way. After explaining the new subject, he moves on to examples and ends with exercises. The exercises are designed or prepared to "deal with" a particular mathematical content, in which the data needed for the solution is already included, without any concern for relating the subject (as far as possible) to situations in the student's daily life. As students try to assimilate the new content and solve the activity, if they don't get the questions right, they turn to the teacher to help them, showing them how to work through the questions.

Thus, in all teaching situations, the teacher tries to convey to the student what he wants him to do. Thus, the reciprocal obligations take the form of "agreements", which Brousseau explains:

> - The teacher is expected to create sufficient conditions for the appropriation of knowledge and to "recognise" this appropriation when it occurs;
> - The student is expected to be able to fulfil these conditions;
> - The teaching relationship must continue, whatever the cost;
> - The teacher therefore ensures that previous acquisitions and new conditions give the pupil the possibility of acquisition (BROUSSEAU, 1996 p. 52).

A well-known example, which can provide a better understanding of this silent adherence to the didactic relationship, is illustrated in the book by psychologist Stella

Baruk: "The age of the captain", published in 1980 in Grenoble (BARUK, 1985 apud D'AMORE, 2006, p.103). Baruck's study analyses the answers given by French secondary school students to the following problem: "On a boat, there are 12 rams and 6 sailors. How old is the captain?" Among the answers given by the students were the following: 18 years old, 72 years old, 36 years old and 144 years old. The mathematical strategies employed by the students (their implicit models) were, respectively, 12+6, 12 x 6, 6^2 and 12^2 (SILVA et al., 1996, p.12). This situation, which has been widely discussed, elucidates the way in which the contract can be established in teaching/learning relationships and reveals that there are specific patterns of behaviour in the pedagogical relationship which are accepted as such by the teaching tradition. The student, seeing himself as a participant in the didactic contract, knows that his role when faced with a mathematical problem is to solve it, without much questioning, without doubting the verisimilitude of the situation, and having to present a number as the result, however absurd it may appear, such as considering 144 years to be a plausible answer to the problem. A situation like this sheds light on how deeply established this didactic contract is, in which everyone assumes their role, without thinking, questioning or trying to break away from this situation.

As with other types of contract, when the parties behave in a way that is not expected, or when something "gets out of hand", it is time to make each party's role clear and, realising that what was being done was not foreseen, there is what Brousseau calls a breach of contract. In this situation, the student must personally assume the rupture of the didactic contract in order to be able to answer that the problem cannot be solved. Once the student assumes this stance, he or she is "freed" from the situation and realises that a new contract is needed from now on. It's up to the teacher to be attentive, to realise that there is this rupture, and to negotiate another type of contract, since the previous one has broken down. It is therefore possible to consider that making the rules of a contract explicit, even if they have not been negotiated, is a moment of awareness, when some doubt, mistrust and/or discomfort arises in relation to a learning situation. It is at this moment, when a contract is broken, that a new contract must be established.

> The contractual rupture would be a creative moment in the pedagogical relationship, both because of the opportunity to regain awareness - on the part of the teacher and the students - of the norms that are at stake in the construction of school knowledge, and because of the possibility of reformulating the ways in which knowledge is appropriated (SILVA et al., 1996, p. 14).

Not only do we consider the contractual rupture to be a moment of resumption, but we also see a certain vulnerability in this moment. If the teacher doesn't intervene, mobilising to establish a new contract and providing a new dynamic, there could be a definitive rupture, generating a disconnection on the part of the student from the didactic situation, which prevents learning.

Pais (2002) gives some examples of ruptures in the didactic contract: the first example of a rupture is when the student doesn't get involved in the activities proposed by the teacher, showing disinterest and apathy; the second example occurs when the teacher proposes solving a problem for which the solution strategy isn't compatible with the student's intellectual and cognitive level; the third and final example of a rupture in the contract is when the teacher shows a pedagogical attitude that isn't compatible with their role of guiding learning situations, letting things get out of hand. One example of a breach of contract is when the teacher applies tests with the aim of punishing the students. These examples of breaches of contract alert us to the teacher's caution and watchful eye in order to notice any kind of breach and act quickly and assertively.

> If learning has not taken place satisfactorily, the work must be redirected to promote a return appropriate to the student's cognitive level. Failure to do so constitutes a breach of contract and implies a withdrawal from the teaching process, and therefore an abandonment of the professional aspect of teaching (PAIS, 2002, p.82).

The teacher needs to be clear about the concept of the didactic contract and recognise it as a "thermometer", which measures the intensity and provides the clarity that the educator must have of the situation in order to act at a given moment and in the right way.

One of the issues that is considered essential and that can guarantee a link between the learning object and the student's reality is the way in which knowledge is presented in a context that gives the student real meaning. The degree of student involvement can be greater or lesser depending on how the different learning activities are structured. According to Brousseau, in relation to a didactic situation:

> [...] everything he [the teacher] does to make the pupil produce the behaviour he expects of him, tends to deprive the latter [the pupil] of the conditions necessary for understanding and learning the notion in question; if the teacher says what he wants, he fails to obtain it (BROSSEAU 1996, p.66, emphasis added).

By this definition, we realise the importance of interactions in this web. The student's interaction with other students, the student's interaction with the object of knowledge, the student's interaction with the teacher and the student's interaction with him/herself, implicitly, in an attempt to give meaning to what is new, to what is being learnt. We must also note that, according to Brousseau, every didactic situation is governed by a certain type of didactic contract, in other words, a didactic situation is formed by the multiple pedagogical relationships established between the students, the teacher and the knowledge. These three component elements of a didactic situation form the essential part of characterising the living space of a classroom. If a situation within the school space lacks one of these components, it will not be a didactic situation. On the other hand, these three elements alone are not enough to "encompass all the complexity of the cognitive phenomenon. Hence the link we make between these situations and other elements of the didactic system: objectives, methods, theoretical positions, didactic resources, among others" (PAIS, 2002, p. 66).

We stress that when developing educational practices, it is important and necessary to be aware of establishing priorities when planning teaching situations, one of which is selecting the content that makes up school programmes. The original source of all this content, which we can also call "school knowledge", is scientific knowledge. We know that there is a process through which this knowledge is selected and transformed into objects to be taught, and this process is carried out by the teacher. The idea of transposition studies the selection that takes place through an extensive network of influences, involving various segments of the educational system. We looked for the definition of "didactic transposition" in Chevallard:

> A piece of knowledge, having been designated as knowledge to be taught, then undergoes a set of adaptive transformations that will make it suitable to take its place among the objects of teaching. The work that turns an object of knowledge to be taught into an object of teaching is called didactic transposition (CHEVALLARD, 1991 apud PAIS, 2002, p. 19).

When we refer to the trajectory of knowledge, we consider the beginning of this knowledge to be determined by a diverse source of influences, which Chevallard calls the noosphere. The noosphere is made up of scientists, teachers, specialists, politicians, book authors and other educational agents. The result of this network of influences brought about by the noosphere has repercussions throughout the teaching system. Selective work results not only in the choice of content, but also in the definition of values, objectives and methods that drive the education system.

We understand that this trajectory through which scientific knowledge travels, until it becomes school knowledge, comes from the noosphere, modifying itself, adapting and, when it is placed entirely in the hands of the teacher, it still runs the risk of transforming itself again, because, from the teacher's point of view, according to D'Amore,

> didactic transposition implies preparing, designing your own lessons, taking knowledge from the source, taking into account the guidelines provided by the instructions and programmes (knowledge to teach) to adapt them to your class, the level of the students, the objectives pursued". (D'AMORE, 2007, p.226).

At this stage of their work, the teacher is not an isolated individual, but a collective, constituted by the educational institution, which in turn has its own objectives and defines school knowledge and methods in its specificity. However, it is through the teacher that content that has undergone a process of selection and various transformations will be transmitted until it becomes school knowledge. The choice of school content is made mainly by the indications contained in curriculum parameters, textbooks, programmes, educational software, material produced by the institution itself, among other sources. However, some content is a true didactic creation incorporated into programmes motivated by supposed teaching needs, serving as a resource to facilitate learning. In an objective analysis, we can consider these didactic creations to have a justifiable educational purpose. The problem can arise when they are used in an automated way, disconnected from their main purpose.

Pais (2002) warns of the permanent care that must prevail throughout the analysis of didactic transposition, because it is the set of didactic creations that highlights the difference between scientific knowledge and taught knowledge. The teacher, acting critically, knows that he or she doesn't have a say in all the stages of didactic transposition, because the choice of the curriculum of the school in which he or she works often doesn't pass through his or her hands. But when they are given the opportunity to give their opinion, make choices and take decisions, be it when choosing a textbook, preparing materials, etc., it is important that they participate effectively and with the necessary rigour, so as not to rely too heavily on didactic creations, resulting in distortions, reductions and a trivialisation of knowledge. It is therefore necessary for teachers to have a grasp of what didactic transposition is, as well as their role and importance in this context.

The teacher, aware of this process, will be able to act in a more active and vigilant way, because it is through their interpretation that they will be able to detect the differences and adaptations that occur between the origin of a maths concept, as found in textbooks, the teacher's teaching intention and, finally, the results obtained in the classroom.

In the quest to understand the theory of didactic situations proposed by Guy Brousseau, we now turn our attention to understanding the didactic contract and its implications for didactic phenomena. This close look is necessary in order to understand and better characterise the relationships and didactic situations that occurred in the geometry sessions, mediated by the Cabri-Géomètre software. Based on Brousseau's theory, we believe that the different forms of a didactic situation are based on the teacher's intention and commitment in conducting this process. However, didactic situations (action, formulation, validation and institutionalisation) geared towards a teaching and learning process are not always permeated by effective and relevant didactic contracts for students to learn mathematical content well.

CHAPTER 3

GEOMETRY IN PRIMARY SCHOOL

This chapter looks at geometry in Brazil, its history and current situation in the Brazilian school context, especially in the final grades of primary school. To this end, the National Curriculum Parameters (PCNs) for the final grades of primary school will be presented first, focusing on the different ways of working suggested for teaching geometry.

Secondly, some recent research on geometry in Brazil will be highlighted, signalling a revival and rediscovery of the subject in schools.

Finally, the importance of geometry will be emphasised, as well as the need to give this field of mathematics its due.

3.1 GEOMETRY IN THE NATIONAL CURRICULUM PARAMETERS (PCNS) FOR THE FINAL GRADES OF PRIMARY SCHOOL

The National Curriculum Parameters (PCNs) provide educators with suggested guidelines for all subjects at the different levels of basic education. In the PCNs for Maths in Primary Education, in the 3rd cycle (BRASIL, 1998), we can see that care is taken to insist that the teacher takes advantage of the reasonable baggage of concepts that the student brings with them, diagnosing how far they have mastered these contents, with a view to their progression in learning. As well as covering the compulsory content of the grade, the document suggests that the teacher take advantage of opportune moments to develop the students' critical sense, giving them the opportunity to get to know the scope of geometry, and also "discussing issues relating to the usefulness of maths, how it was constructed, how it can contribute to solving both everyday problems and problems linked to scientific research" (BRASIL, 1998, p. 62).

When we analyse the PCNs, we notice the concern with geometry, especially the emphasis given to the relationship between geometric entities, their visualisation and their use in solving problem situations. Among the objectives of the PCNs, in relation to geometric thinking, the parameters propose:

> Solving problem situations involving the location and displacement of points in space, recognising in the notions of direction and direction, angle, parallelism and perpendicularism fundamental elements for the constitution of Cartesian coordinate systems; establishing relationships between spatial figures and their plane representations, involving the observation of figures from different points of view, constructing and interpreting their representations; solving problem situations involving plane geometric figures, using decomposition and composition, transformation, enlargement and reduction procedures (BRASIL, 1998, p. 64 - 65).

With these objectives for the 3rd cycle, the contents are proposed in such a way as to cover them in their entirety. For the 4th cycle, the concern with visualisation and interpretation is maintained, and the emphasis is given in a similar way to the 3rd cycle, but according to this level, respecting the sequential and gradual form of the themes, including already glimpsing the maturity to be reached.

There are recommendations that maths learning should be anchored in social contexts that clearly show the relationship between mathematical knowledge and work (BRASIL, 1998). The objectives (linked to geometric thinking) proposed for the 4th cycle are:

> Interpreting and representing the location and displacement of a figure on the Cartesian plane; producing and analysing transformations and enlargements/reductions of plane geometric figures, identifying their variant and invariant elements, developing the concept of congruence and similarity; broadening and deepening geometric notions such as incidence, parallelism, perpendicularism and angle in order to establish relationships, including metric ones, in two- and three-dimensional figures (BRASIL, 1999, p. 81-82).

Based on these objectives, we realise that there are topics and themes aimed at treating geometry, as well as a concern to relate it to other areas of mathematics.

In addition to the links that should be established between maths and cross-cutting themes, there is an emphasis on developing autonomy, a spirit of enquiry and the ability to solve problems. The PCNs emphasise that one of the objectives for primary school is to develop students' ability to use different sources of information and technological resources to construct concepts. These considerations indicate that the parameters suggest consistent goals to be achieved at this level of education. In addition, other indications suggested by the parameters consist of opting for problem solving as the starting point for mathematical activity, including due emphasis on the history of maths and communication technologies.

For teachers, understanding these guidelines and then reflecting on them can provoke a series of questions, such as: "How can we achieve these goals?". Reading and understanding this document awakens educators to the need to "enforce" their practice, adding to the content the development of skills in line with society's needs.

The PCNs provide a historical context, stating that in the 1960s and 1970s, in Brazil and other countries, there was a movement that influenced maths teaching, the so-called Modern Maths Movement[3] . He stresses that "it was born as an educational movement within a policy of economic modernisation [...] which, together with the area of Science, constituted a privileged access route for scientific and technological thinking" (BRASIL, 1998, p. 19). This innovative proposal had as its axis

[3] For references to the Modern Maths Movement, the acronym MMM will be used.

the Theory of Sets, integrated elements of algebraic, topological and logical structures. The MMM provoked discussions and reforms in the maths curriculum worldwide.

When the PCNs cite the Modern Maths Movement as one of the factors that contributed to the abandonment of geometry in schools, the justification for this fact is that the MMM was gradually weakened because its proposal was beyond the reach of students and also because of the lack of understanding of the proposal on the part of teachers at the time. We believe that this movement influenced the trajectory of geometry in this context, which can be seen in the conclusions of Pavanello's (1989) study, when he states that the teaching of geometry gradually disappeared from the school curriculum and was neglected over the years, with geometry often being scheduled for the end of the year. According to the author, this fact worries many teachers. In her master's thesis, Pavanello (1989) points to some historical and social factors.

The social factor cited by the author is that "the problem of geometry arises and grows as secondary schools begin to cater for an increasing number of students from the less favoured classes" (PAVANELLO, 1989, p.180). Geometry then goes through a process of exclusion from the curriculum, and in some more restricted cases, it is worked on in a much more formal way, starting with the introduction of Modern Mathematics. According to the author, this moment happens "just when the struggle for the democratisation of educational opportunities is intensifying" (PAVANELLO, 1989, p. 180).

The historical factor for the abandonment of geometry is presented in the form of a historical review, in which the author chooses significant periods and works that could indicate the paths taken by geometry. Pavanello (1989) discusses the political, social and economic aspects of Brazilian society, pointing out the realities of inequalities between classes. According to Pavanello (1989), before 1931 the majority of the population still had no access to education, not even elementary education. Maths content was taught separately (arithmetic, algebra and geometry) by different teachers. There was a discriminatory character between the classes, with the less favoured classes having different opportunities from the more favoured classes, as the following excerpt shows:

> In any case, the duality of the education offered to the elites and the masses is clear. [...]. It is also expressed in the objective with which the various subjects are taught in each type of school - which is quite significant from the point of view of the teaching of Geometry (PAVANELLO, 1989, p. 87).

One of the sources used by the author was Law 5692/71, the Law of Guidelines and Bases for 1st and 2nd Grade Education. In 1975, the São Paulo State Secretariat produced and published the "Maths Curriculum Guide", which made the following recommendations, among others:

> - a course in intuitive geometry for the first four grades;
> - a study of measurements, which is <u>much more appropriate and more likely</u>

> to be assimilated in a science course;
> - the study of geometry in the 5th grade[a] as a vehicle for introducing the language of set theory;
> - the introduction of geometry through transformations from the 7th grade onwards (PAVANELLO, 1989, p. 164, emphasis added).

These resolutions make up the list of reasons why geometry has been neglected in favour of algebra, as the author states that many teachers, because they did not have the necessary mastery to work with geometry focusing on transformations, stopped teaching it. In addition, when analysing Law 5692/71, the author states that this law allowed each teacher to draw up their own syllabus, "according to the needs of the clientele" (PAVANELLO, 1989, p. 165). Considering that the above resolutions date back to 1975, it is possible to understand the "deformations" to which the teaching of geometry has been subjected, thus creating a vicious cycle in which the teacher, without mastery, does not work with geometry, and geometry, because it is not worked on, does not require teachers to prepare and/or train and loses its importance. As it loses its importance, it no longer becomes compulsory and therefore loses its place in the school curriculum.

The author concludes that the teaching of geometry was weakened not only by the introduction of the MMM, but that this problem also increased as secondary schools catered for students from the less favoured classes.

> Geometry gradually disappeared from the school curriculum with the introduction of Modern Mathematics, which took place just as the struggle for the democratisation of educational opportunities intensified, concomitant with the need to expand schooling to a more significant portion of the population (PAVANELLO, 1989, p.180).

The work by Pereira (2001) entitled "A geometria escolar: uma análise dos estudos sobre o abandono de seu ensino" also indicates that geometry is being abandoned. This work analysed theses and dissertations that studied the causes of the abandonment of school geometry, providing an overview of the research carried out on this subject. In her conclusion, the author points to the Modern Maths Movement as an indicator of the "decline" of Geometry teaching. About this movement, the author states:

> In the midst of the need for renewal, the Modern Maths Movement was basically shaken by the attempt to once again unify the three fundamental branches of maths: sets, relations and structures (PEREIRA, 2001, p.64).

The author suggests a proposal to re-qualify teachers as an attempt to reverse this situation.

In this proposal, there are records in the collections of the Programme for Study and Research in the Teaching of Mathematics (PROEM) of results in the form of interviews with 903 teachers, most of whom say: "In Geometry, there are important contents, but not essential ones" (PROEM, p. 45, v. 2.1 apud PEREIRA, 2001, p. 65).

In addition to the Modern Maths Movement, the author emphasises two main aspects: problems with teacher training and the omission of geometry from textbooks.

According to Zuin (2006), primary school curricula in Brazil underwent changes with the enactment of Law 5692/71. There was a list of compulsory subjects and another list of optional subjects. Schools could choose which optional subject to integrate into the curriculum. One of the stipulations of the school legislation was that Art Education should be included in all grades of primary and secondary education. Geometric Drawing had become an optional subject and this led many schools to abolish the teaching of geometric constructions, which had once been the syllabus for this subject. Another factor cited by Zuin (2006, p.1) is that "geometric constructions with rulers and compasses would no longer be compulsory in Architecture and Engineering entrance exams in the 1970s". These facts prove the "abandonment" of Geometric Drawing in the final grades of primary school.

Recent research carried out in Brazil on maths in schools, and more specifically on the teaching of geometry, points to some causes for its decline and the fact that it has less and less space in curricula, putting it at a disadvantage compared to other areas of maths.

From the research mentioned below, we can see that most of them address the situation of geometry in Brazil, which is below expectations, and point to the computer, if used with a constructivist, interactive and dynamic vision, as an alternative to reduce this gap The research was carried out in different contexts, some with primary school students as research subjects, others analysing teachers. The research presented has as its "common thread" the problems surrounding the teaching and learning of geometry and the use of technological resources.

Bertoluci (2003), whose work aims to evaluate the influence of the use of software on the learning of geometric concepts, analysed 24 5th graders[a] from a public school in Jaú. The author points out the serious difficulties in teaching geometry in Brazil in recent decades and considers Cabri-Géomètre as an alternative for overcoming these difficulties. The author concludes that the software helps in understanding geometric concepts, but emphasises the urgent need to invest in training teachers to be able to develop activities with the computer.

Zullato (2002), whose research aimed to analyse the profile of teachers who use Dynamic Geometry (DG) software, had 15 teachers from different states (SP, RJ, DF) as the subjects of his research. The study concluded, among other things, that when teachers have a profile of researching, investigating and reflecting on their practice, they go in search of alternatives, such as technological tools. The teachers reported two main difficulties in using computers: too little equipment for too many students and a lack of computer maintenance, which causes technical problems. The theme of this research is linked to continuing teacher training, which is a never-ending path of research and investigation in order to constantly mature and innovate teaching practice. Zulatto (2002) also observes that teachers use few demonstrations. The teachers who took part in the research themselves claimed that in recent years students have been lagging behind in content, which makes them immature enough to follow the development of a demonstration.

Fonseca (2001) proposed a study of geometry in primary school using interactive learning resources. The research subjects were students in the 8th grade[a] in Belo Horizonte. The research also signalled the abandonment of geometry teaching, pointing to poor teacher training and the exaggerated importance of textbooks as factors contributing to this reality. The author carried out a comparative analysis between the traditional teaching method and the use of Cabri software. The research concluded that the Cabri-Géomètre II software provided the perfect conditions for teachers to use modern, up-to-date teaching methods, and showed that, for the proposed teaching of geometry in the 8th grade of primary school, it left nothing to be desired compared to traditional methods.

The studies analysed reveal a scenario of concern and uncertainty regarding the direction of geometry in schools. The authors' justifications and conclusions for the lack of knowledge of geometry, which has been neglected in most schools, have generated a contingent of students who are unable to make the simplest distinctions between flat and spatial figures. For many students, a square, a flat figure, can mean a cube, a spatial figure, and the opposite is also true. These gaps presented by many students can lead to problems in the development of spatial and dimensional vision, as I understand geometry to be a tool for understanding, describing and interacting with the space and world in which we live.

3.2 NEW SCENARIOS FOR THE TREATMENT OF GEOMETRY

Even if we have data that allows us to conclude that geometry has been treated less than "deserved", there is, on the other hand, research pointing to a scenario of rediscovery and inclusion of geometry in schools.

One indicator of this is a survey carried out by a group of researchers, GT 19, who are part of Anped (National Association for Research in Education)[4] . The research professors are divided into research groups by area and sub-area.

The research presented at the Anped WG[5] , with a historical-bibliographical approach,

[4] ANPED aims to develop and consolidate postgraduate teaching and research in the field of education in Brazil. Over the years, it has established itself in the country and beyond as an important forum for debating scientific and political issues in the field, and has become a reference for monitoring Brazilian production in the field of education.

[5] This research is entitled Didactic-pedagogical trends for the teaching of Geometry, whose authors are:

studied the Annals of the National Maths Education Meetings (ENEM). Its aim was to identify and analyse current didactic-pedagogical trends in the teaching of geometry in Brazil, from 1987 to 2001, when the first seven meetings were held. This work also sought to identify researchers and/or research groups working in this area, as well as the theoretical and epistemological assumptions that have been subsidising these discussions.

The guiding question of this research was: "What didactic-pedagogical trends are present in the teaching of Geometry, taking the Annals of the National Maths Education Meetings (ENEM) as a reference?" The study reports on the varied research on the subject of geometry, totalling 363 papers, and the credibility of these papers is due to the fact that, for the group, the ENEMs constituted the "highest level of discussion and circulation of academic productions in the area, at least until 2000, when the International Seminar on Research in Mathematics Education (SIPEM) was created" (ANDRADE, NACARATO, 2004, p.1). The group's analysis initially identified seven categories for the teaching of geometry. Of these, two became the object of analysis because they were considered to be "emerging didactic-pedagogical trends": Experimental geometry and Geometry in computer environments. According to the study (2004, p.3), the works included in the category called Geometry in Computational Environments (23% of works) were organised into three subcategories:

> a)Geometry in Dynamic Geometry Environments (Cabri-Géométre, Geometricks, Geometer's Sketchpad, Tabulae and Mangaba);
>
> b)Geometry in the LOGO Environment;
>
> c)Other.

The work points out that the themes developed on Dynamic Geometry environments have shown the influence of French Mathematics Didactics and identified that "some of the concepts from this line of research also subsidise theoretical discussions relating to the teaching of Geometry using Cabri-Geomètre". The most cited theories in this line are: Vergnaud's theory, Douady's theory and Duval's theory.

One aspect that was shown in the works that use French Maths Didactics is that the authors don't just use one concept from this line of research, as is the case with Didactic Engineering[6] , which was present in several works.

If the intention is to situate geometry in the research landscape, this work from Anped's WG19 details and places geometry as a current object of study and also portrays a new approach to it. The subcategories of the study allow us to realise that there are numerous works on the Cabri-

---

ANDRADE, José Antonio Araújo - USF/SPNACARATO, Adair Mendes- USF/SP. The work was presented at the 27ª meeting of ANPED, from 21 to 24 November 2004, in Caxambu.

[6]Didactic engineering is a particular way of organising the methodological procedures of research into the didactics of mathematics (PAIS, 2002, p.99).

Géomètre software[7] , also from the perspective of French mathematics:

> The concepts of French Maths Didactics are also present in research using the Cabri-Geomètre computer environment. This research initially appeared at the V ENEM (3 papers), but the big "explosion" occurred at the VI ENEM (16 papers) and VII ENEM (22 papers) (ANDRADE; NACARATO, 2004, p.13).

The data from this research indicates this new approach that research, production and investigation into geometry has focused on. We can understand from these trends in research that the need to rethink the teaching of geometry under a new approach is a current concern for educators and mathematicians. As a conclusion to the WG19 research, the group states:

> We believe that the movement that has taken place in the production of works on the teaching of Geometry is related to the very constitution of the Maths Education research community. The resumption of the teaching of geometry in basic education will require new studies and theoretical discussions; these, in turn, will make it possible to expand the lines of research in this area (ANDRADE; NACARATO, 2004, p.16).

Other work also reveals the need to fill in the gaps and wake up to the problems surrounding the teaching of geometry in schools.

3.3 WHY WE SHOULD TEACH GEOMETRY

Geometry is considered an auxiliary tool for understanding, describing and interrelating with the space in which we live. Without an understanding of the notions of space and shapes, we will be restricted, "stuck", with difficulties in developing geometric thinking and the formulation of visual reasoning, thus generating difficulties in solving real situations, be it the notion of space, area, or even calculation with estimates. Without a knowledge of geometry, their interpretative view of the world becomes incomplete and reduced.

Geometry is everywhere, you just have to be attentive and have a slightly stimulated geometric vision to realise it. What would art be without geometry? Even if we are reluctant to see it, in everyday life we deal with notions of space, shape, perpendicularity, parallelism, similarity, congruence, symmetry, proportionality, measurement (area, volume, length). Geometry manifests itself in the most varied forms, and in the various sectors of our lives, whether in leisure, profession or oral communication.

Research into maths, with specific reference to geometry, highlights the need to teach geometry. In his master's thesis, Silva (2006) points out that, in addition to the reasons mentioned

[7] For the Cabri-Géomètre software, only the first name will be used, i.e. just Cabri.

above, which show us the importance and need to work with geometry, there are many others that are implicit, as he states:

> The purpose of teaching geometry is not limited to solving practical problems. Approaching a way of reasoning without relying exclusively on the perceptible is the greatest motivation in making the decision to work with school geometry (SILVA, 2006a, p.136).

In her work, this author concluded that the approach to geometry in the early grades is currently dependent on teachers who do not have mathematical training, along with the fact that the teachers who were the subjects of the research "expressed uncertainty about teaching the contents of the Space and Shape axis" (SILVA, 2006a, p.136). Conclusions such as these, resulting from work with primary school teachers, give us a sense of how real and equally serious the problem involving geometry is, since the interviews conducted with these teachers indicated that they do not consider geometric content (from the space and shape axis) to be fundamental, either because they do not know it, and/or because they find it difficult to teach it.

Fainguelernt (1999) says that geometry needs to be taught in schools, but questions the way in which it is approached, as she does not agree that the teaching of this field of maths should be reduced to the application of formulas and results established by theorems. She says that teaching geometry is justified by the need to discover ways of demonstrating and deducing its formulas, without the need to stick to the exhaustive process of formalisation. Another difficulty is the role of demonstrations, together with the relationships between intuition, induction, deduction and rigour. One aspect of this problem lies in knowing when these levels of abstraction should be explored, considering the age and consequent maturity of the students. The author believes that there is a lack of integration between some of the elements that make up the teaching and learning of geometry:

> The contents of Arithmetic, Algebra, Analysis and Geometry, considered as if they were totally watertight; the actions of the teachers in the classroom and their own objectives; the results of the students' performance in carrying out the activities and the teachers' reflection on their praxis (FAINGUELERNT, 1999, p. 23).

The topics raised above lead to the opposite reasoning: if we were able to combat them, we would be able to guarantee the effectiveness of geometry teaching and learning. The author, believing that Maths Education has been on the rise on the educational scene, either because of its research or because of its relevance and close relationship with classroom issues, defends a new scenario that is taking shape, concomitant with the advances and facilities that technology offers. A new approach to teaching through the use of computers, in the sense of "reducing some learning difficulties, enabling the creation of a space for the exploration and construction of knowledge [...] giving a dynamic dimension to this teaching" (FAINGUELERNT, 1999, p.15).

There is no doubt that it is necessary to rethink the teaching of geometry in schools. However, it will do little good if the attempt is to teach geometry that is disconnected from reality, static and meaningless. In order to avoid working with geometry that is out of context, we sought a way of linking it to the new, the dynamic, the movement, enabling students to visualise and manipulate the components of this geometry.

The focus of the study carried out here in relation to geometry is certainly what happens here in Brazil, but authors from other countries have written and reported on the same problems faced in that country. Usiskin (1998), from the University of Chicago, in his article entitled "Resolving the enduring dilemmas of school geometry", points out two main problems raised by work on geometry: (a) poor student performance and (b) an outdated curriculum. These conclusions are not dissimilar to the situation here in Brazil. According to the author, students don't learn geometry because their teacher doesn't know how to teach it (this teacher certainly won't emphasise the presence of geometry around the student). On the other hand, if the teacher doesn't know how to teach, it's because he didn't learn it in his degree course either. Most likely, when that student becomes a teacher, they won't teach it either. The question that arises is: how can this situation be overcome and resolved?

The author offers suggestions for solving problems, emphasising that this is not an easy or immediate task, but one that must be started. Amongst others, he suggests

- The need to specify a geometry curriculum for primary schools by grade, requiring all students to have a significant degree of competence in geometry;
- Make the subject of geometry compulsory in universities, not just optional;
- Discuss the purposes of geometry more clearly;
- To analyse the various ways of forming concepts in geometry, from a curricular perspective (USISKIN, 1998, p.24).

These conclusions and ways of resolving these dilemmas in school geometry also apply to our problems with school geometry, making us realise that these problems do not only occur here in Brazil, and that the issues surrounding the problems are very similar, thus showing us that their suggestions should be considered and reflected upon.

Usiskin (1998) also considers other issues relating to geometry, such as its approach in the curriculum, and assigns four dimensions to the study of geometry:

> a)Geometry as the study of visualisation, drawing and the construction of figures (the measurement-visualisation dimension);
> b)Geometry as a study of the real, physical world (the real-physical world dimension); c)Geometry as a vehicle for representing other mathematical concepts (the representation dimension);
> d) Geometry as an example of a mathematical system (the mathematical

> support dimension) (USISKIN, 1998, p.34).

Two other dimensions emerge, which are represented in school maths: a socio-cultural dimension and a cognitive dimension of understanding. For the author, observing and treating geometry in terms of these dimensions would be the start of a new geometry, as he states below:

> Geometry requires drawing simple figures and interpreting visual models. These models interact continuously with the physical world, with other parts of mathematics and can be logically interrelated in various ways (USISKIN, 1998, p. 35).

This reflection on geometry that the author provides is directly linked to the problems experienced in the classroom, and the dimensions listed above lead us to rethink the teaching of geometry, its difficulties and the inevitable path towards alternatives that deal with it. The suggestions made by the author are at the heart of geometry teaching, suggesting that the solutions to its problems lie in geometry itself. However, this research uses a slightly different approach to those proposed by Usiskin (1998), seeking help and support from technologies.

Through the use of computers, we believe it is possible to find alternatives to re-signify the teaching of geometry. Through the use of appropriate software, which allows us to give movement and dynamism to the exploration of geometric concepts, we can configure a path that is directly linked to the teacher-student relationships implicit in a teaching and learning situation.

We end this chapter having addressed some of the issues that permeate geometry. We have presented work that has pointed to geometry being abandoned and the factors that have led to this situation. We show works that indicate a revival of research that puts geometry at the centre of study and re-signification. We also discuss the importance of understanding geometry and the need to take it seriously and work with it in the classroom.

Next, we will focus on an element that is believed to be an aid in the revival of this geometry that has been sidelined: the computer, more specifically the geometry software called Cabri-Géomètre.

CHAPTER 4

CABRI-GÉOMÈTRE SOFTWARE

From now on, we will look at issues relating to technological advances and the implications of using computers in school practices. Next, Cabri-Geomètre, the technological resource focussed on in the object of study, will be presented. Firstly, the arrival of computers in Brazil on the school scene will be described, citing the countries that influenced the introduction of this resource. With regard to Cabri, its origins, its use and some of the possibilities offered by the software will be presented. In addition, current studies will be pointed out that recognise this software as a suitable tool for dealing with geometry. After discussing Cabri, we will end by emphasising the importance of the teacher in this scenario of using such a support tool.

4.1 A SUPPORT TOOL: THE COMPUTER / EDUCATION IN THE FACE OF TECHNOLOGICAL ADVANCES

One of the ways of overcoming the problems detected as agents of failure in geometry points to the use of computers in this context. According to Valente (1998), the arrival of computers in schools provoked questions, the predominant issue being whether the computer had the "power" to replace the teacher in the classroom. Other forms of reaction to this have emerged, which the author calls visions, divided into sceptical, indifferent and optimistic.

The first group (the sceptics) thought that computers were just another fad and that there would be no way of implementing them in schools because of the financial costs, as well as the fact that computers are rational, cold machines that could dehumanise schools. The 2nd group (the indifferent) didn't define themselves, showing apathy and waiting for the facts to evolve. The 3rd group (optimists) believed that the computer would come to repair existing gaps in education.

We can easily see that the information revolution has influenced all sectors of society. It is even present in those sectors where it is least expected. This revolution has provoked in individuals the need to acquire new skills, as there is no longer any room for routine and repetitive qualifications. It's no different in the school environment, when we see the way students communicate, with terms from the internet influencing oral and written language (internetese). In the research and school work carried out by students, the main source of research is the internet. Therefore, in the school context, the changes become even more decisive and important, because it is from the school that society as a whole will be transformed.

In relation to the use of the Internet as an educational tool, Kalinke (2003) points out that there are various resources and possibilities that are seen as positive aspects of this global computer network in educational processes, but emphasises that the greatest highlight of the use of the Internet is due to "the interaction it allows, whether between students, between students and teachers or between students and machines, the ease of communication, the possibility of publishing materials and the ease of access to information". (KALINKE, 2003, p. 42) Here we point to the Internet as an educational tool, emphasising that there are various ways of using the computer, in addition to using educational software. Of course, in order for the teacher to use the Internet in the classroom, it is essential to screen it, which Kalinke (2003) calls "analysis criteria". For the author, there are two main aspects to guide the criteria: constructivist aspects and ergonomic aspects[8] . Teachers can choose how to incorporate the computer into their classroom practice, be it via the internet, educational software, simulations, using Excel, etc. What is clear is that this trend can no longer be avoided.

Convinced that there is no escaping this reality, schools need to be aware of this trend, which is advancing rapidly and has taken hold for good. The computer, when used as a teaching tool, appears here as a great possibility for changing a real situation of disarticulation between subjects, starting from the fact that the content learnt in the classroom remains watertight and the student's reality is far removed from the school context. Access to the media is somewhat more widespread, and in some state schools you can find computer labs, giving students contact with this technology and allowing them a greater scope in the process of socialising knowledge. Within this proposal, the teacher plays an important role, mediating between the knowledge made available by the computer tool and the student. In order to play their true role, it is up to the educator to accept and credit the computer with being a catalyst for profound change, which is considered a true paradigm shift, starting with the teacher's attitude towards this new scenario.

Valente (1998) uses nomenclatures for the paradigms that permeate teaching with computers: the author sees them as antagonistic, and that one of them must overcome the other. They are the instructionist and constructionist paradigms.

4.1.1 The Instructionist paradigm

The instructionist paradigm, by placing the computer as a teaching machine, insists on the computerisation of traditional teaching methods. Instead of using the computer in a way that favours meaningful, investigative and dynamic learning, in this paradigm, the teacher, out of ignorance, lack

[8] According to Kalinke (2003), ergonomics on an educational website can be seen as essential for learning to take place. To cover the criteria, the author suggests looking at whether the site has good readability, provides documentation and has good navigability (KALINKE, 2003, p. 117).

of preparation or convenience, uses it as a tutorial, in which the student learns by simply following instructions. In this approach, the teacher remains the transmitter of knowledge and the student remains passive, not learning with any meaning, but following the machine's instructions, repeating exercises without questioning. And what is a real obstacle to professional development is that many teachers, by using the computer in some class in the way mentioned above, feel that they are innovators, adopting a mistakenly modern stance. For Papert (1994), the term instructionalism is far from being discussed in pedagogical terms, as instructionalism "should be read in a more ideological or programmatic term as expressing the belief that the way to better learning must be through the improvement of instruction" (PAPERT, 1994, p.124).

4.1.2 The constructionist paradigm

The constructionist paradigm opposes the previous paradigm in terms of computer use. Papert (1994)[9] was the one who called the use of computers for real pedagogical purposes, as an aid in the construction of knowledge, constructionism. Papert used the term constructionism to show another level of knowledge construction, where knowledge construction is based on carrying out a concrete action that results in a tangible product, developed with the help of the computer. For the author,: knowledge construction takes place when the student constructs an object of interest, such as a work of art, an experience report or a computer programme.

> In Papert's notion of constructionism, there are two ideas that make this type of knowledge construction different from Piaget's constructivism. Firstly, the learner is building something, in other words, learning by doing, by "getting their hands dirty". Secondly, the learner is building something that interests them and for which they are highly motivated. Affective involvement makes learning more meaningful (VALENTE, 1998, p. 40).

In view of these names for the use of computers, we would emphasise that there is really no justification for using them for instructional purposes, under a purely traditional approach. However, it's not always, or rather not all the time, that students are offered total freedom to "create" an object of their interest. Firstly, there are school regulations, and each educational institution has a Political Pedagogical Project that must be respected; secondly, there is a curriculum that includes certain content for each grade; thirdly, respecting the specificities of each subject, in maths, it is not always that the student can create something of interest. We should also remember that school practices in general do not follow this purely constructivist characteristic.

We believe that it is a challenge for teachers, in addition to the use of computers in their

[9] Seymour Papert is a professor and creator of the LOGO language. He worked with Jean Piaget, from where the basis for the constructionist paradigm, which incorporates the idea of "construction" elaborated by Piaget, emerged.

practice, to create/develop situations where students can feel motivated, challenged to discover and unravel this new object of knowledge in a creative way.

Valente (1998) considers that what most differentiates constructionism from constructivism is the presence of the computer: "The fact that the learner is constructing something through the computer (computer as tool)". For the author, using the tool creatively to get the most out of learning requires certain actions that are very effective in the learning process:

> When learners interact with computers, they are manipulating concepts and this contributes to their mental development. They are acquiring concepts in the same way that they acquire concepts when they interact with objects in the world, as Piaget observed. Papert called this type of learning "Piagetian learning" (PAPERT, 1980 apud VALENTE1998, p.55.).

We think it would be good to clarify these nomenclatures and their definitions, emphasising that the teacher can look for ways to use the computer under a constructivist approach, respecting the specificities of the subject and taking advantage of opportunities to do so in a way that allows the student to construct knowledge.

4.2 COMPUTERISATION IN SCHOOLS: A SLOW AND DEFINITIVE JOURNEY

If today we are no longer amazed by a classroom using a whiteboard with a multimedia projector suspended from the ceiling, a computer connected to the Internet, with a USB stick and sound input, etc., perhaps we should look back five or ten years. If described years ago, today's scenario could intimidate many teachers - which doesn't mean that it can still intimidate many others - who still have the blackboard, white chalk and their textbook as the only sources of interaction with their students, apart from themselves.

If we consider that it is now possible for teachers to prepare their lessons with animation, sounds, images, films, slides, graphic diagrams, simulations and, in the classroom, to open a web page and share it with their students, their horizons will expand infinitely. These options available to teachers in the face of technological resources can offer them some support, motivating them to incorporate them into their practice once and for all. The term "definitive" doesn't really fit when it comes to technology, because discoveries, inventions and new versions mean that a device, a resource, becomes obsolete very quickly, just as the information and data released about an event is updated.

This evolution has taken place in all sectors, be it industry, science or education. The arrival of computers in schools cannot be treated in isolation from the technological changes that have taken place in the world over the last 30 years. These changes, more specifically in the financial and industrial sectors, have begun to have repercussions within schools, given the need to train new professionals who can meet these new demands, i.e. a new type of professional, with a new profile, focused on technological advances and their applicability in the various sectors.

In Brazil, the use of computers in education began with some experiments at universities in the early 1970s. According to Valente (1999), several universities began using computers for certain (and different) purposes. One of the motivating factors behind the idea of implementing computer use in schools came from technological movements in countries such as the United States and France.

However, in both France and the United States, there were doubts about the real role of computers in schools. Should universities and schools teach students how to handle the computer, learn programming languages, or should it be used as a support for the student? This doubt gives an idea of the scale of the uncertainties that accompanied the novelty of computers in schools. To better delineate this situation, Valente explains the arrival of computers in schools in the United States:

> The use of computers in schools is pressurised by technological development and the competition established by the free market of software companies, universities and schools. Technological changes are fantastic and tangible, but they don't correspond to pedagogical changes (VALENTE, 1999, p.3).

The US also saw the development of programmed instruction methods using the computer, based on the use of the computer as a "teaching machine", devised by Skinner in the early 1970s. This system of programmed instruction was called CAI: Computer-Assisted Instruction, produced by companies such as IBM, RCA and Digital, and was used mainly in universities. With the introduction of microcomputers in schools in the 1980s, CAIs spread widely, with some being modified and others being created, such as tutorials, demonstration programmes, drill-and-practice, learning assessment, educational games and simulation. However, according to Valente (1999), this number of options for using the computer made it possible to publicise new educational possibilities, such as helping to solve problems, produce texts and so on. From this perspective, the computer was seen as an important auxiliary tool, complementing and "possibly changing the quality of education", making it possible to create learning environments, such as the LOGO language[10] , for example. However, the proposal to use LOGO was not fully accepted, as teachers were not properly trained and this resource was under-utilised.

Today, with the help of educational software and/or activities that make use of the internet, computers are being used for some pedagogical purposes and can become moments of extreme richness and significant progress in terms of learning, but the emphasis is still on computer-based instruction.

Sancho (2006, p.19) takes a stand on the evolution of computers in schools, stating that

---

[10] LOGO is a programming language developed by Papert in 1967, based on Piaget's theory and some ideas from Artificial Intelligence.

"many people interested in education have seen digital information and communication technologies as a new determinant, a new opportunity to rethink and improve education." The author goes further, saying :

> The recent history of education is full of broken promises and unfulfilled expectations, generated by each wave of technological production (from the paperback book to the video or the computer itself) (SANCHO2006 ,p.19 apud SANCHO,1998).

One of the problems why the arrival of computers in schools has been slow and less than the creators believed it to be is pointed out by Sancho (2006, p.19) when he refers to the dominant teaching model in schools, centred on the teacher.

According to Valente (1999), American universities are still the main trainers of teachers in the area of information technology, but the training/preparation for education professionals still trains them to work with information technology in the sense of transmitting information.

For Area (2006), the 1990s represented a period in the international context of critical review of what had been achieved in relation to the use of computers in schools, in other words, "neither was it demonstrated that students learnt more and better, [...] nor did teachers in general innovate their practices by adopting computers as a regular teaching resource; the expected pedagogical revolution did not take place" (AREA, 2006, p.155).

When discussing the evolution of the use of computers in schools in France, Valente (1999) states that the country took greater care in training teachers. The implementation of computers in French schools was planned in terms of the target audience, materials, distribution methods, installation and maintenance of equipment in schools; the French government was also concerned with the phase that precedes the use of computers with students, in other words, the "teacher-intelligence", dedicating at least a year of preparation for the teacher in front of the computer, before using it in the classroom. This preoccupation with preparing teachers demonstrates the country's vision of education, i.e. a certain caution in adopting a new stance, even though it is aware of the need to do so. "Although the aim of introducing IT into education in France has not been to bring about pedagogical changes, it is possible to see progress in this direction, but this progress is far from the desired transformations" (VALENTE, 1999, p.36).

Properly pedagogical computer training began in France in four stages, from 1970 until the "IT for All Plan" in 1985, when there was a greater proliferation of computers in schools. Teacher training programmes were developed in the thousands to cater for millions of pupils; another concern of the French programme has been to guarantee all individuals access to information and the use of information technology. This has now been reinforced by projects to set up computer networks and distance communication for education and training.

The way computers were brought into schools and promoted did not follow a universal model, and therefore did not occur in the same way in each country. This is only natural, as each country had its own objectives, interests and government policies involved. There are also different authors' visions of the countries' objectives when introducing this tool.

Valente (1999) believes that since the introduction of computers, changes have not been very significant in either country. Referring to France, the author says: "The content was about the study of computer science and computers, as well as an introduction to programming languages, without establishing links between educational theories and pedagogical practices with computers".

Oliveira (1997) points out that in both the United States and France, the aim was the same: "to improve the quality of schools and guarantee students access to knowledge of a technology that is extremely widely used in modern societies". Even with different authors' views on the subject, the fact is that computers arrived in schools and this attitude "branched out" to other countries and continents, arriving here in Brazil.

If the ways in which computers were introduced in each country had their own purpose and evolution, Brazil also had its own way, considering the external influences, mainly from the United States and France.

According to Oliveira (1998), the arrival of computers in schools was not the decision of educators and education activists, but the will of the Brazilian government, which felt it was necessary to involve public schools in a movement that was already taking place in developed countries: technological evolution.

In Brazil, the first governmental actions aimed at linking information technology with education took place in 1979, when the government created Capre[11] , later replaced by SEI[11 12] , whose purpose was to regulate, supervise and encourage the development and technological transition of the information technology sector.

In 1980, SEI set up the Special Education Commission, with the responsibility of gathering input to generate standards and guidelines for the area of information technology in education. The idea was to start a process that aimed to link education with information technology. However, in practice it didn't seem to happen, because according to Oliveira (1998) until 1980 "the IT-education binomial only occupied space within the state bureaucracy, since there was no interference from sectors of society directly linked to education". (OLIVEIRA, 1998, p. 29)

With the intention of involving the educational community in this discussion, the implementation of the IT in Education programme in Brazil began with the first and second National

---

[11] Commission for the Coordination of Electronic Processing Activities. Government organisation that coordinated initiatives in the field of IT.

[12] Special Secretariat for Information Technology, directly linked to the National Security Council (CSN).

Seminars on IT in Education, held respectively at the University of Brasilia in 1981 and the Federal University of Bahia in 1982. These seminars concluded that the computer needed to be seen as an auxiliary resource for developing skills, in other words, a facilitator. These seminars established a programme of action that gave rise to EDUCOM - Education with Computers. This project represented the first official and concrete action to bring computers into Brazilian public schools. Educom's main objective was to stimulate the development of multidisciplinary research, focusing on the application/use of computer technologies in the teaching-learning process, carried out through pilot projects.

The above is intended to situate the arrival of computerisation processes in schools on an "outside-in" scale. It was possible to see that there was an influence from the government, which had some interest in the implementation and effective use of this "machine". Analysing developed countries, Brazilian governments understood that the way forward was to invest in this technology. We can see that the evolution took place gradually, through trial and error, with pilot projects developed by some universities. Today many schools have computers inside them, but the use of this tool certainly falls short of the proposals and intentions for this purpose.

4.3 THE CURRENT COMPUTER LANDSCAPE: ALLY OR NUISANCE?

The path that led to the arrival of computers in schools, through the Educom project, had its needs and, with that, its implications, as well as its ramifications. These include the Formar project, aimed at training public school teachers, and the Centre for Information Technology in Education (CIED) project, aimed at primary and secondary schools in the public education system.

It was only natural that these needs arose within schools, because the arrival of computers in schools did not mean having them exclusively for bureaucratic use in the secretariat or treasury. This premise of having computers in the classroom meant first of all a change in teacher training, and then it also meant preparing the physical space for the equipment, as well as solutions relating to its maintenance. Faced with so many changes, how did this machine reach the student?

To start answering this question, all we have to do is look back and ask ourselves if we were students in the 1980s. If the answer is yes, we will have some story or memory of that time. There's no doubt that students were more receptive to computers than teachers, because expectations were enormous! If we're talking about public schools, even more so, as the school was seen as the only way to access this universe. How anxious they were to see the equipment being installed in the schools, eagerly awaiting the day when they could tinker with it, explore it and discover so many other things with it!

However, when we analyse the current reality, we realise that the initial proposal, which bet on the "revolution" that the arrival of computers in schools would promote, has not happened and has frustrated many educators. In an attempt to answer this question, Sancho (2006) says: "the recent history of education is full of broken promises; of unfulfilled expectations, generated by each new wave of technological production (from the paperback book to the video or the computer itself)" (SANCHO, 1998 apud SANCHO, 2006, p. 19).

There are many reasons for concluding that computers are inefficient or underused in schools, but Sancho (2006) attributes this to the fact that teaching is still teacher-centred. We believe that teacher training is the fundamental foundation for improving the quality of teaching, and we know that such training is defined by political, economic, cultural, ideological and pedagogical elements. So perhaps the gap between the way computers are used in schools and the various approaches that could be used is what Sancho 2006 points out: "In an increasingly complex society, attempts to situate students' learning and their educational needs in the school of pedagogical action are still in the minority" (CUBAN, 1993 apud SANCHO 2006, p.19). As a result of this gap, there is no going backwards, but rather forwards towards overcoming this obstacle. Teachers, aware of this reality, must look for ways to bring about pedagogical and technological change in schools.

4.4 COMPUTER SCIENCE'S CONTRIBUTIONS TO THE TEACHING OF GEOMETRY

This study has described the teaching of geometry, and we have seen that there is a gap in its approach. We also reported above that the use of computers for teaching purposes is less than expected. Given this scenario, we see the need to implement actions of various kinds for the teaching of maths and, more specifically, geometry.

One of the alternatives we present as a possibility for revitalising and improving the teaching of geometry is to include the use of new technologies in teaching practices. This can be done through software, programming languages (LOGO), among other possibilities.

Some authors point the way, such as Papert (1994) through the LOGO language and constructionist theory, and Fainguelernt (1999) with learning geometry through LOGO. There are also authors who advocate the use of Dynamic Geometry, such as Gravina (1998); Rodrigues and Braviano (2002), Zulatto (2002), among others.

More specifically in the use of software as an aid to learning geometry, there is the category of Dynamic Geometry (DG) software. According to Zulatto (2002), the term Dynamic Geometry was originally used by Nick Jacw and Steve Rasmossen in a generic way, seeking to differentiate Dynamic Geometry software from other Geometry software. What characterises this difference? Dynamic Geometry software[13] (DG) has a feature that offers the option of dragging, allowing continuous transformation of the drawing or construction in real time.

GD software provides resources with which students can carry out geometric constructions that until then had been done with rulers and compasses, taking up more time. In this context,

[13] For references to the term Dynamic Geometry, the acronym GD will be used.

exploration activities are developed in which students can check, validate and redo their constructions by interacting with the computer.

Gravina and Santarosa (1998) describe some of the characteristics of programmes built on the principles of Dynamic Geometry:

> They are construction tools: drawing objects and geometric configurations based on the properties that define them. Through displacements applied to the elements that make up the drawing, it is transformed, maintaining the geometric relationships that characterise the situation (GRAVINA; SANTAROSA, 1998, p. 14).

The authors (1998) present three characteristics that should be among the necessary requirements for learning, when referring to computerised environments within a constructivist proposal. These are:

1- The environments must be dynamic;
2- The environments must be interactive;
3- The environments must be conducive to modelling and simulation.

When they refer to the dynamic medium, the authors want to emphasise the counterpoint to the static character that has been given to systems for representing mathematical knowledge. In this way, a mathematical object is given a changeable character, providing dynamism through direct manipulation of the representations that appear on the computer screen.

The interactive media mentioned by the authors go far beyond the types in which the machine interacts with the user only to inform them of successes or errors. The interactivity expected of a constructivist learning environment is when the system supports the students' realisations and mental actions, allowing the student to act, in other words, making it possible to manipulate what is represented on the computer screen. There is also another feature offered in this proposal, which is the capturing of procedures, in which all the student's procedures are recorded and, whenever necessary, it is possible to go through the whole process of their construction. This is a special feature of geometry programmes.

In modelling or simulation, students build models and/or carry out experiments of a more complex nature, such as population growth simulations, in which students qualitatively explore mathematical relationships, enabling them to see mathematics as a tool for solving problems in other areas of knowledge.

The authors present a piece of software as an example of an environment that has the characteristics described above: Cabri-Géomètre. We believe that this device is capable of becoming an alternative for combining geometry with the use of computers, as it is dynamic geometry software.

4.5 CABRI-GÉOMÈTRE SOFTWARE

Cabri-Géomètre was developed at the Joseph Fourier University in Grenoble, France. The acronym Cabri comes from the French, Cahier de Brouillon Informatique, which means Computerised Sketchbook. A group of specialists worked on this software for four years under the coordination of Jean Marie Laborde and Frank Bellemain (BONGIOVANNI, 1997). Nicolas Balacheff, Jérome Borde, Gerard Cossavella, Mireille Dupraz, Michel Guillerault, Bachir Kskessa, Colette Laborde, Gerard Lejeune, Michel Mollard, Charles Payan, Salima Tabri and Gerard Vivier collaborated on the development of the software and documentation (Cabri Manual 1991).

In Brazil, Cabri-Géomètre was marketed exclusively by the Centre for Exact Sciences and Technology at PUC-SP in the early 1990s.

The Cabri-Géomètre software[14] consists of an interactive drawing book that offers alternatives for teaching geometry, allowing experimentation, observation and analysis of constructed geometric figures. With it you can draw geometric figures, measure segments and angles, determine geometric locations, among many other functions.

> The software allows you to build all the geometrical figures you can draw on a sheet of paper with a ruler and compass, with the huge advantage that the set of these primitives can be expanded thanks to the "macro construction" feature, which allows you to define a new construction by designating its results or final elements, starting from a prototype on the screen (CABRI MANUAL, 1991, p.10).

4.6 CABRI-GEOMETRE IN BRAZIL

According to the Cabri Manual[15] (1991), the software is an appropriate tool for learning geometry, offering a learning environment in a microworld. The idea of a microworld implies the existence of an environment made available to the user, in this case the student, to carry out experiments, exploring a particular universe and discovering its properties in an investigative way.

In the table below, we present a short history of Cabri, following its trajectory and arrival in Brazil.

Year	**Occurrence**
1981 a 1985	Work on Cabri-Géomètre in graph theory.
1985	Informal specifications for creating a computerised sketchbook.
1986	Cabri-Géomètre prototype in three doctoral theses.
1987	Pre-product and in-class experimentation.
1988	Apple Trophy for best teaching achievement; I^{a} public demonstration (ICME Budapest).

[14] References to Cabri-Géomètre will most often be referred to as Cabri.

[15] The Cabri software is available on floppy disc and CD-ROM media. To install it, the computer must have the following configuration: 486/Windows 3.1/8MB RAM/ SVGA/Mouse.

1989	First edition of Cabri-Géomètre in France.
1990	Enabling the IMAG Cabri-Géomètre project.
1992	Creation of the Cabri-Géomètre international research group
1993	0 Cabri-Géomètre is translated into 25 languages and commercialised in 40 countries. In Brazil, the software is commercialised by PUC-SP

Chart 1 - Chronology of Cabri-Géomètre Source: Bongiovanni (1997, p. 6)

According to the table presented, we can see the progress and acceptance of Cabri on a global scale, with good acceptance and also the commercialisation of Cabri in Brazil.

The first version of Cabri was presented in 1988 at the International Congress of Mathematics Education in Budapest. Since then there have been changes and new versions of the software. A1ª version used was Cabri I, then Cabri II (which is the one used in this work). There are other versions of the software, called Cabri Plus and Cabri 3D respectively. On the Cabri website[16] and on software commercialisation sites, there is a Demo version available for download[17] , but this version, once "downloaded", lasts about fifteen minutes of use, i.e. just for a quick experiment. Cabri is currently in its third version (Cabri II Plus) and, although development continues under the care of the IMAG (Grenoble Institute of Computer Science and Applied Mathematics), it is commercialised by Texas Instruments, both for PCs and for some calculator models. More recently, an old concept has been adapted for Dynamic Geometry. 3D computer modelling has already been widely used in a wide variety of areas. The Cabri team developed Cabri 3D, which applies 3D modelling to the study of spatial geometry.

From the works, articles and dissertations that refer to this software, we can see that it is accepted by the students, and the teachers who use it see it as an appropriate support for maths content.

It should be emphasised that for reasons of availability we chose the Cabri software, the focus of this investigation, in the knowledge that it is not the only software used to teach geometry, but that it is one of the most pedagogical and interactive. Even though we are familiar with other software, we believe that Cabri offers the user a clear, attractive and easy-to-use interface.

4.7 THE CABRI-GÉOMÈTRE ENVIRONMENT

Cabri-Géomètre was developed to allow students to explore the world of elementary geometry. When students drew geometric figures on a sheet of paper, they had a more or less precise representation, but it was always immobile. With Cabri, figures can be described by actions and in a language very close to that of the familiar "pencil-paper" universe, offering the user "electronic rulers and compasses", and the interface of construction menus in geometric language.

The Cabri toolbar contains options that allow you to generate constructions. There are 11

[16] The Cabri website can be found at the following address: <www.cabri.com.br>.

[17] Downloading a programme means making it available for your use by transferring files from a remote computer to your own.

toolboxes, as shown in Figure 2:

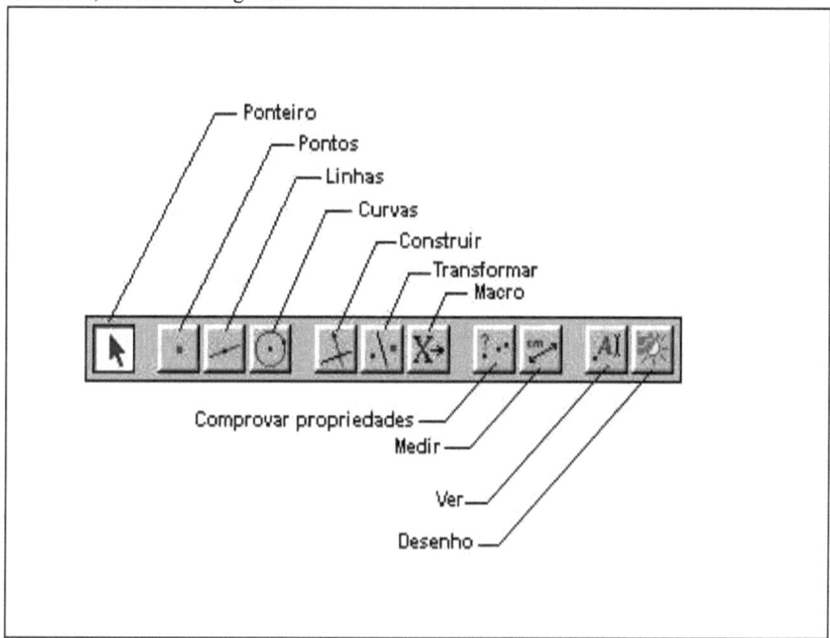

Figure 2 - Cabri-Géomètre Software Toolbar Source: Cabri Manual, 1991

The toolbar in Figure 2 is what appears when the user opens the screen, i.e. the interface that the software presents. To give you a better idea, Figure 3 shows the same toolbar and its proportion in relation to the Cabri screen.

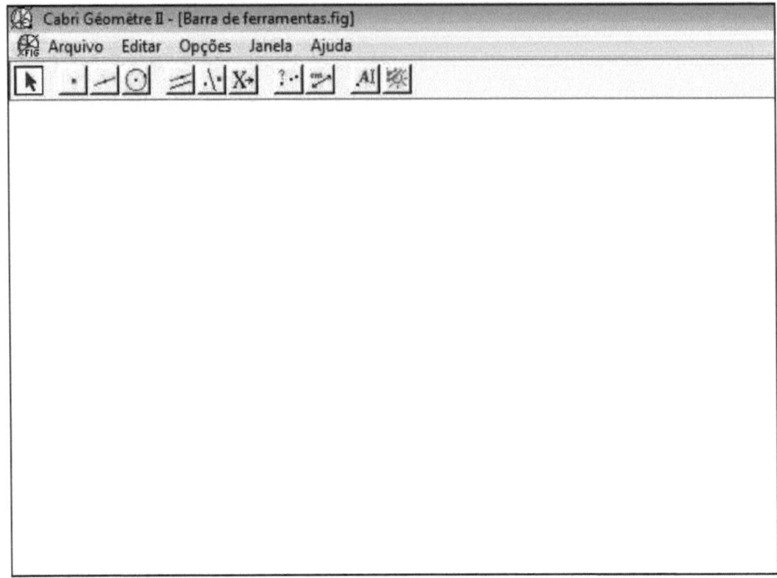

Figure 3: Cabri screen Source: The author

Figures, when drawn, can be deformed directly on the screen by moving their base elements, while still retaining the properties assigned to them. More precisely, the initial geometric points of a construction can be dragged with the mouse without destroying the mathematical relationships between them and the other objects. In this way, we can study the same construction for different configurations of points without having to repeat the construction. This is illustrated in the figures below. Any triangle (Figure 4-A) shows the measures of its internal angles, with the sum next to it. When we move the vertices of the triangles, the figure "deforms" (Figure 4-B), but the properties of the sum of the interior angles remain the same.

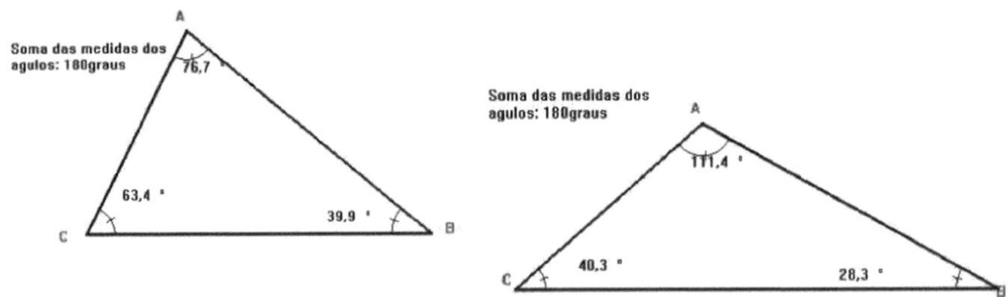

Figure 4 - A - Any triangle
Figure 4-B - Any triangle after moving
Source: The author

This software option, which allows you to "drag" elements on the screen, whether it's the whole figure or some of its objects, falls under the heading of Dynamic Geometry software.

As well as being able to drag, the drawings are constructed based on the properties that define them. So, to delete an object drawn on the screen (point, line segment, circle), the student simply clicks on the respective object, or can use the eraser option. However, care must be taken in this process, as all objects constructed from the object to be deleted will also be deleted.

In the context of learning with a dynamic geometry environment, the teacher, when using the software to construct geometric objects, gives the student the freedom to move the elements that make up the drawing, allowing it to be transformed and still maintain the same geometric relationships that characterise the drawn figure. In this way, when we observe an object on the screen, we have associated with it a collection of "moving drawings" and the invariant characteristics that appear there correspond to the geometric properties of the object in question.

We insisted in this explanation on the resources that the software offers so that the user can construct any and all geometric figures, beyond the use of rulers and compasses. With this, they can go further, i.e. the set of constructions can be extended by a feature called "macro-construction" which allows them to store a new construction from a prototype which, under the "macro" command, will appear on the screen.

When referring to what is possible with the use of this tool, Silva (1997, p.48) states that "developing an activity in Cabri means using geometric relationships and properties, even if the content developed is not geometric. In other words, you have to see the concept in the geometric field." In this way, the same concept can be explored in different ways, providing a broader understanding of a given topic.

Learning geometry from this perspective is different from using a ruler and compass, with visible differences. And no attention will be paid to comparisons of "what you gain (or lose)" by using this tool. It is therefore up to the teacher to have the perception to choose the moment when there is the necessary maturity so that the use of Cabri is not, in reality, an under-utilisation.

According to the tutorial that the software offers, some important commands and menus can be highlighted. The features of this software tutorial are listed below:

Chart 2 - Cabri's main features

Features:
It includes analytic, transformational and Euclidean geometry;
It allows intuitive construction of points, lines, triangles, polygons, circles and other basic objects;
Moves, enlarges (or reduces) and rotates geometric objects relative to their geometric centres or specific points;
Easily construct conics, including ellipses and hyperbolas;
Explores advanced concepts of descriptive and hyperbolic geometry;
Notes and measures the figures (with automatic updating);
It uses Cartesian and polar coordinates;

It provides a presentation of the equations of geometric objects, including lines, circles, ellipses and coordinates of points;
It allows users to build macros for frequently repeated constructions;
It allows the teacher to configure the tool menus to focus on the students' activities.

Source: Cabri Manual

From the table above, it was possible to see the range of options on offer, as well as others, which will depend on how much (or little) the teacher explores and proposes suitable activities. The tool offers many options, but in order to use it effectively and try to exploit much of what is on offer, the teacher first needs to be interested in learning how to use it.

using the software, its menus and commands. Then, when the teacher has mastered the tool, their moment of greatest autonomy takes place: studying and deciding which concepts can be worked on and how these concepts will be worked on. In this way, the teacher is often induced to recall basic geometry concepts in order to analyse and prioritise the best way to use this tool, with the aim of providing the learner with an exciting, dynamic and interactive approach.

Interaction with the user takes place through menus and commands, which form the basis of the user interface. The menus are divided into

- Editing: cancelling, deleting everything, naming, etc.
- Creation: point, line, circle, etc.
- Construction: geometric place, point on object, mediatrix, etc.
- Miscellaneous: removing an object, macro construction, measuring, marking angles, etc.

In addition to plane geometry, the software allows you to develop activities with other maths topics such as:

- Algebra: working with 1st and 2nd degree polynomials and constructing graphs;
- Trigonometry: exploring the trigonometric cycle;
- Spatial Geometry: representing some geometric solids on the plane using cavalier perspective;
- Analytical geometry: allowing you to calculate distances, obtain the equations of geometric objects (lines, circles, ellipses and the coordinates of points); construct parabolas, ellipses, hyperbolas, etc.

To give you a clearer idea of what the software interface looks like, Figure 5 shows the construction of a regular polygon inscribed in a circle, and some measurements.

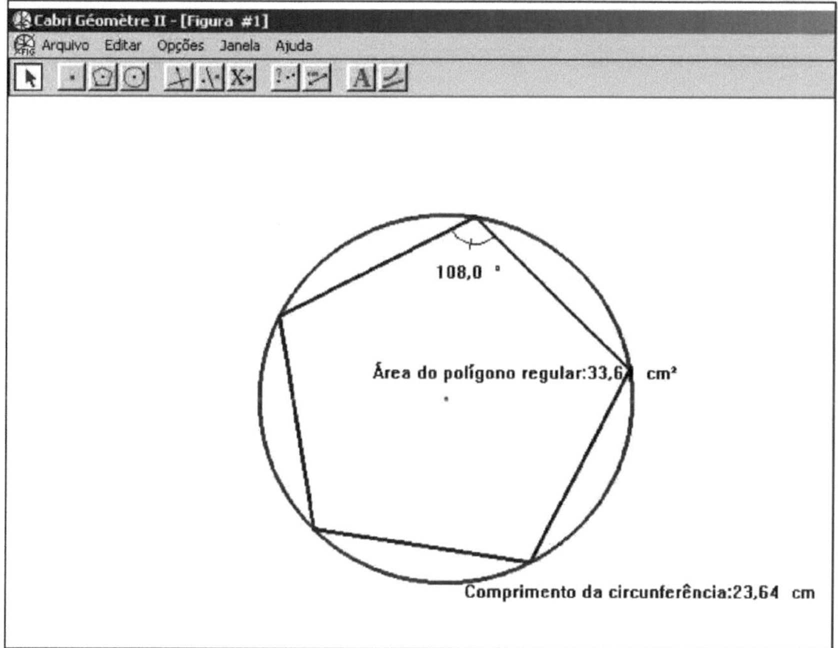

Figure 5 - Regular polygon built in Cabri with measurements of sides and angles Source: The author

It is important to emphasise that software alone cannot guarantee or bring definitive solutions to the formation of geometric concepts. Any change will only occur if the teacher is willing to explore and incorporate the new concept into their practice.

Software is presented in the teaching and learning context as a tool, and tools only carry out tasks under human supervision/guidance. In this case, with the use of software, the teacher's action is fundamental if the work developed with the student is to be productive, so that concepts can be constructed and validated.

4.8 THE IMPORTANCE OF CABRI-GÉOMÈTRE FOR TEACHING AND LEARNING GEOMETRY

The use of Cabri-Géomètre software, which is the subject of this study, meets the various questions that have always permeated teaching practice. Reflecting on the experiences in professional practice, we realised that there are numerous difficulties in teaching geometry, such as:

a) geometry occupies little space in the curriculum;
b) students find it very difficult to understand geometric concepts;

c) geometry lessons offer few opportunities for students to handle instruments such as rulers, compasses and protractors.

This perception of the discrediting of geometry in school curricula is disturbing, because we recognise the importance of this subject for students' mathematical education, especially for the construction of spatial and geometric notions, since these notions are present in problems related to the shapes of the physical world and in the development of deductive mathematical thinking.

When referring to geometry, Fainguelernt (1999, p. 21) states that "in recent years, research into geometry has been widely stimulated by new ideas from mathematics itself and from other disciplines, including computer science". In this statement, the intention to combine the aspects of exploration, visualisation and manipulation of geometric entities that Cabri software can offer to the study of geometry is evident.

As explained above, Cabri is software whose basic tools are geometric; however, it allows you to carry out activities that are not necessarily geometric. Developing an activity in Cabri means using geometric relationships and properties, even if the content developed is not geometric. We can construct function graphs, obtain the equation of the line on the coordinate axis, and other options, but we need to see the geometric relationship involved. Therefore, a given topic (a straight line, for example) can be explored in different ways, with different approaches, thus enabling a greater understanding of the chosen subject.

Figure 6 shows a Cartesian coordinate axis and a line on it. If the content in question is not the equation of the line, the teacher can still explore the line in the quadrants, move it around and observe the sign of the coefficients. He can also ask the student to construct other lines parallel and/or perpendicular to the existing one.

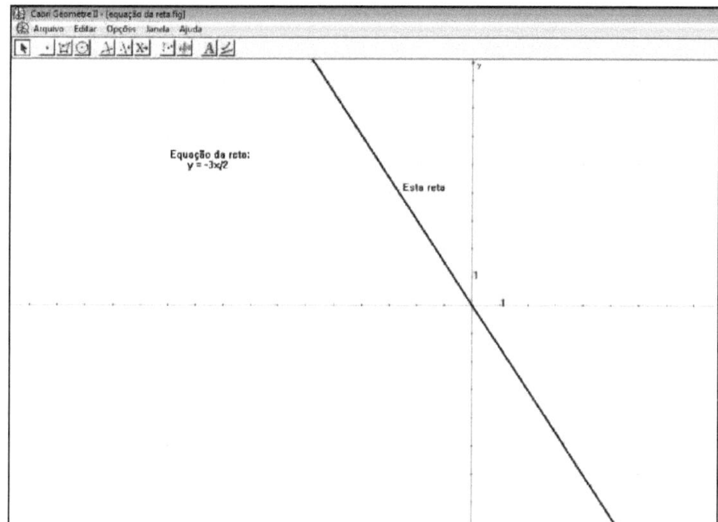

Figure 6 - Line on a Cartesian coordinate axis.
Source: The author

The example above illustrates the construction and possible exploration of a line on a coordinate axis, but it is possible to use Cabri to explore a variety of concepts in addition to basic geometric concepts, such as trigonometry (trigonometric cycle), spatial geometry (perspective of spatial figures), descriptive geometry (épura) and art education.

The interface that the software offers, without any user intervention, is a blank screen, requiring the student to explore the buttons and menus, to move the mouse in order for anything to happen. Thus, "in order to analyse a given situation, we must first construct it" (BONGIOVANNI, 1997, p. 3).

Other authors who deal with this tool emphasise the different characteristics of Cabri, presenting its potential and/or limitations. Gravina and Santarosa (1998), referring to the software, say: "It is a tool especially for geometry constructions. It has an "electronic ruler and compass", and the construction menu interface is in classic Geometry language (GRAVINA; SANTAROSA 1998, p.14)."

To summarise, it is important to clarify that in order to analyse the didactic contracts that are established in a computer-mediated lesson (in the form of Cabri), it is necessary to understand the possibilities offered by the software. More than that, it is necessary to understand the process the student goes through to construct the figures,

At this point, we mixed the necessary previous content with the new themes, as well as handling the software's menus. We are aware of the subtleties and specificities that make up the vertices of the triangle formed by the elements: Cabri, computer and geometric knowledge, linked to the student. Figure 7 illustrates this situation. We are aware of the need to sharpen our gaze among the multiple interactions that will take place. It is essential to sharpen our gaze among the multiple interactions that will take place. We need to be aware of the fact that geometric knowledge, including Cabri and the computer, are intertwined, even though they have different functions and even though the student interacts with these elements in different ways.

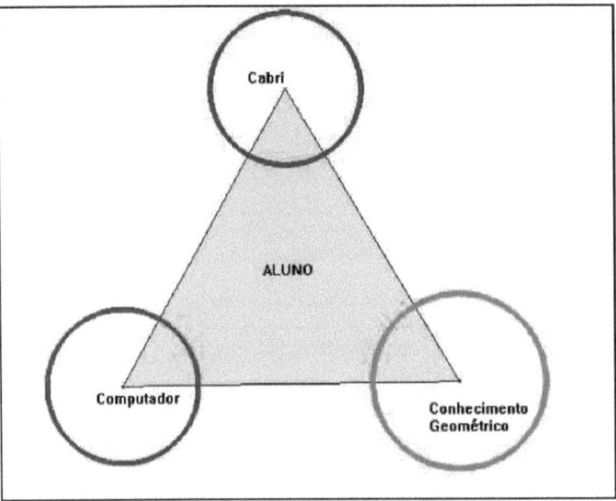

Figure 7 - Student interactions with the elements: Cabri, computer and geometric knowledge.
Source: The author

CHAPTER 5

METHODOLOGICAL ASPECTS OF THE RESEARCH

In this chapter we try to detail the methodological elements that guided this investigation into the use of Cabri-Géomètre, characterising the approach chosen, the universe in which the data was collected, the subjects involved, the procedures used, in short, a breakdown of the path taken and the techniques used to understand the object of study.

As explained in the introduction to this paper, our concern as a maths teacher and our interest in using Cabri-Géomètre were due to our perception of the students' lack of preparation in relation to geometry, in particular our concern about the fact that the students did not master the basic concepts of geometry, nor were they able to differentiate between the main characteristics of a square and a rectangle, or the confusion and lack of knowledge between quadrilaterals.

As we got to know the software, we became even more curious, and why not say we needed to understand what it could mean in terms of gains in learning geometry. However, we also realised that excitement alone would not be enough to bring about significant advances in teaching practice or any improvement in learning geometry. We needed to understand the macro context of this situation, and in a theoretically grounded way, so that we could realise our aspirations and expectations.

In this sense, we believe that trying to understand the didactic relationships that occur when learning geometry using a technological tool would be an opportunity to reflect on our own experience of teaching and learning geometry. The question that guided this study was in the context of the geometry teaching and learning process: with the Cabri Géomètre resource, how are didactic relationships established between teacher, student and mathematical knowledge?

In order to carry out the research, we opted for a qualitative approach, looking to authors such as Bogdan; Bikle (1994), Ludke; André (1986), among others, to understand the characteristics of this approach, with a view to making good use of the data collection and analysis procedures.

5.1 RESEARCH WITH A QUALITATIVE APPROACH

In using the term qualitative approach for this research, we considered what Bogdan and Biklen (1994, p.47) say when referring to the characteristics of qualitative research:

1. In qualitative research, the direct source of data is the natural environment, and the researcher is the main instrument. Qualitative researchers go to the places of study because they care about the context; however much equipment is used to collect data (filming, audio, etc.), there is a need to complement it with information obtained through direct contact, i.e. video and/or audio alone

are not sufficient elements for data collection, but mainly the perception of the researcher, with their inquisitive and attentive attitude.

2. Qualitative research is descriptive. For a broader view of the research universe to be possible, the richness of the description of the environment, the subjects involved and the phenomenon to be observed is vital. Data is collected in the form of words, images or even sounds, but not numbers. The data collected includes: interview transcripts, field notes, photographs, videos, personal documents, memos and other official records.

3. Qualitative researchers are more interested in the process than simply the results or products. While every detail is important, the results will be meaningful if they emerge from a context, with a logic and some coherence in relation to the data presented.

4. Qualitative researchers tend to analyse their data inductively. In order for the researcher to reach a conclusion, they must first construct a scenario that takes shape and becomes more consistent as they collect and analyse the data.

5. Meaning is of vital importance in the qualitative approach. More than analysing the data, it is important for the researcher to try to understand the link between the elements that make up the phenomenon, because according to the authors, "qualitative research sheds light on the internal dynamics of situations, dynamics that are often invisible to the outside observer".

Having explained the characteristics of qualitative research, we would like to emphasise the universe of this research, making clear the changing and dynamic nature of the relationships that are established in a classroom. Since it is necessary to explain the relationship between teacher and student in the context of computer-mediated learning, this research was carried out in a classroom environment using computers, more specifically in a computer laboratory.

The key elements of our observations were: students, teachers, the learning environment and the geometry content of grade 6[a] , together with the Cabri-Géomètre software. Even though we have delimited the object of study, there is a need to develop an inquisitive, investigative and attentive look at the component elements of the phenomenon to be studied. As (LÚDKE; ANDRÉ, 1986, p.22) emphasise: "The importance of determining the foci of the investigation and establishing the contours of the study stems from the fact that it will never be possible to explore all the angles of the phenomenon in a reasonably limited time." In this research, the outlines were characterised in the contracts that were established between teacher, student, geometry and the use of Cabri-Géomètre.

As Ludke and André (1986, p.12) state, "in qualitative research, all the data from reality is considered important". In this sense, the relationships arising from this context became fundamental elements in the construction of the object of study. The research carried out was of the case study type, as it took place in a natural situation, described the facts and the units were made up of sixth-form primary school students and the maths teacher in the class.

The authors Ludke and André (1986) characterise case studies under a qualitative approach:

> We can say that the "qualitative" case study has great potential for getting to know and understand school problems better. By portraying everyday school life in all its richness, this type of research offers precious elements for a better understanding of the role of the school and its relations with other institutions in society (LÚDKE; ANDRÉ, 1986, p.23).

We therefore opted for a case study, using a qualitative approach, to carry out this research.

5.2 RESEARCH UNIVERSE

The school where the research was carried out is part of Curitiba's private education network. It works with its own teaching material, which is produced by the institution's Centre of Excellence, with the participation of some teachers and professionals in the field.

production of the material. It caters for all levels of education, from nursery school to higher education. The institution has three units in Curitiba, as well as working with partner schools throughout Brazil, which are schools that use this institution's teaching materials.

The research was carried out in just one of the units, the one where we work as primary school maths teachers. This unit has two computer labs, one with 26 and the other with 21 computers; four multimedia rooms; a science lab; an art room; and a teaching kitchen. Of all the classrooms, three have the multimedia resource installed in the classrooms themselves, so there is no need for students to travel; the other classes use the specific classroom for multimedia use when necessary, with the need for prior scheduling by the teacher who needs to use it. The school has audio-visual resources such as TV, video and sound, which are taken to the classrooms on request. Primary school classes do not exceed 45 students. All the classrooms have a whiteboard and individual desks.

Most of the students at the institution under investigation are very familiar with computers. Those who have been attending the school for more than a year are already familiar with this resource, as they have been doing activities in the laboratory since the early grades. Most of the students have a computer at home with an Internet connection. This is due to the fact that the institution uses, among other tools, the School Portal, as well as the Eureka virtual environment. Both the School Portal and Eureka are connected via the Internet, and some activities, whether research or assignments, are worked on as an extension of the classroom, so students use these virtual environments in their own homes. The Eureka tool is used systematically at the institution, not just as a "repository" of activities, but as an effective teacher-student communication channel. In this virtual environment, each grade is subdivided into the subjects that make it up, in which the student enters each room (one room per subject) separately, taking part in forums, chats (designed and run by the teacher), visiting new links, sending emails directly to the teacher of the subject. The Interactive School Portal[18] is a website

[18] The interactive school portal can be accessed at: www.escolainterativa.com.br

created and maintained by the school's pedagogical team. The information contained in this portal refers to the school universe, from content divided by grade and subject, to a place for parents to access. In the case of access by parents, a password is provided at the start of the school year so that family members can keep track of bimonthly averages, events taking place at the school, among other options. Among the options offered by the Portal, navigation between topics is easy and it contains hyperlinks that make it possible to navigate between the various topics. As mentioned above, students have been using computers as an aid in teaching activities since the early primary school years, demonstrating that this educational institution emphasises the use of technology in its pedagogical proposal. The school presents this emphasis on technological resources as a differential in relation to other institutions of the same size.

The teachers at this school use the computer as a support tool, constantly taking their students to the computer lab to carry out activities. The Cabri-Géomètre software was acquired by the school 10 years ago and the maths teachers at this unit use it to teach geometry.

With regard to the use of Cabri-Géomètre, most of the students are already familiar with it, as the school uses it from grade 5ª . The school also uses other educational software, such as Imagine[19] , suitable for each level of education. As well as educational software, the students also develop activities with Robotics and Lego Zoom[20] , working on programming the robots and activities linked to the themes developed in the classroom, in the Maths and Science subjects.

The proposal to use Cabri wasn't just about going to the computer lab to use the resource in isolation. In 2007, this class developed an interdisciplinary project involving the arts and maths disciplines that studied Plato's Polyhedra. Theoretical studies and activities carried out in the classroom were complemented by activities with Cabri, and vice versa. The use of Cabri was therefore a complement, helping to understand theoretical-practical relationships. It is already embedded in the school culture of this institution that the activities to be developed in the laboratory must be linked to a macro context, contextualised and in line with its proposal.

5.3 THE SUBJECTS INVOLVED AND THE FIRST CONTACT

The research was carried out with 6ª primary school students. The class investigated consisted of 37 students aged between 11 and 12 in 2007. As mentioned above, the students in this class used computers easily at home and at school, so they were familiar with using the equipment.

These additional characteristics revealed about the research subjects make it possible to perceive the different profile of the students, their familiarity with the computer and contact with the

---

[19] Imagine is authoring software that offers the user many options. Imagine software allows you to work with sounds and images, animation and has the LOGO language feature.

[20] Lego Zoom is made up of educational kits comprising building blocks, construction sheets, gears and the RCX brick. The aim of the activity is to build robots, programmed to fulfil a certain challenge, where the student must find the solution to the proposed situation using the programming language.

programming language, a relevant factor to be considered in the analyses of the data obtained in this investigation, since understanding that this familiarity with the computer, (not necessarily with the software in question) and the skills they have in dealing with the machine were decisive elements in their use of Cabri.

The teacher of the class under investigation has been teaching primary school at the institution for fifteen years. She has a degree in Maths from the Paranaguá State College of Philosophy, Sciences and Letters (FUNFAFI) and has been teaching for 17 years. She is very outgoing and well-liked by her students, fostering a classroom environment permeated with trust, allowing the students to feel close to her. We can say that the teacher of the class under investigation manages to balance her outgoing manner with the energy needed to intervene in moments of conversation and behaviour considered inappropriate for the classroom environment. We can see that the teacher has the ability to promote a dynamic classroom environment without losing sight of the students' learning. We won't go into the affective issue, which can influence learning, but this characteristic is marked in this teacher and we believe that this particularity constitutes a significant aspect in the relationship between teacher x student x knowledge.

5.4 RESEARCH PROCEDURES

The research consisted of six class observation sessions. The lessons were recorded for later transcription and analysis. The observations took place in the computer lab. The teacher conducted the lesson as normal and the recorder was kept in her pocket, so that the students, even though they knew the lesson was being recorded, did not deal with the intimidating presence of the recorder, thus trying to put them at ease as much as possible. In addition to the tape recorder, the researcher recorded the moments in a "logbook", which was made at each observation session. The "logbook" consists of a notebook in which all the events considered relevant to the research phenomenon are recorded. According to Ludke and André (1986), these notes have their reflective part, which includes the researcher's personal observations made during the collection phase, i.e. speculations, feelings, problems, ideas, impressions, preconceptions, doubts and uncertainties. This reflective part is a guide, a guideline to guide the selection of what to observe and also to help organise the data.

Class observation was the main element of the research, as we believe it is the most comprehensive way of capturing and identifying the relationships and interactions in a classroom environment mediated by Cabri, which make up the subject of this research. As mentioned above, observations alone are not enough to collect data, but above all the perception of the researcher, with his inquisitive and attentive attitude.

The observations were then planned in such a way as to define "what" and "how" to observe, starting with the object of study. The focus was on the students and the teacher using the Cabri-Géomètre software. In the "logbook" we tried to make descriptive records, taking care to separate what seemed relevant from what was trivial.

The main advantages according to Ludke and André (1986, p.26) are that, firstly, "the observer can draw on personal knowledge and experience as an aid in the process of understanding and interpreting the phenomenon studied", which confirms the understanding that the degree of involvement of the researcher with the subject and their academic maturity can have a significant influence on the use and abstraction of the facts surrounding the phenomenon. Secondly, direct observation allows the observer to get close to the "subjects' perspective", blending what they observe with their world view. As for the disadvantages, there are some aspects that can be considered problems, or that can distort the environment in which the situation being researched is taking place; one of them is the fact that the observation method can cause changes in the environment or in the behaviour of the people being observed, which is perfectly understandable if you consider the researcher to be a stranger to the environment and the established routine; another criticism is that this method relies too much on personal interpretation, or that too much time spent by the researcher with the subjects can lead to a distorted view of the phenomenon or a partial representation of reality. From what has been said so far, it is understood that the choice of observation can cover the variables surrounding the problem in question.

There were six observations in all, each lasting no more than 50 minutes. That's the time of one lesson, and we also deducted the time spent changing rooms, which lasted about five minutes.

5.5 DESCRIPTION OF THE RESEARCH: THE FIRST CONTACT

The first contact with the students took place in June 2007, in a maths lesson, when the class teacher first took the students to the multimedia room, where she explained the "movement" and the reason for the researcher's presence in the room. We first explained that she was a research teacher and that she was going to carry out a research project in which they would be the "main actors". We explained that we wanted to accompany them in the next lessons and that the research would consist of observing the sessions (lessons) and that the focus of the study would be the relationships that would occur in maths lessons, with themes related to geometry, using the Cabri software. We also talked about the possible contributions of this research. We explained the importance of the research subjects, because without them the research couldn't happen. Many looked at each other, proud of having been chosen, and said that they had never taken part in anything like this before. We opted for the researcher to lead the explanation, along with clarifying why she was teaching them to use the software, and not the class teacher. We explained to the class that this lesson was the beginning of a series of meetings we would be having, and that they would be observed using Cabri. It was also made clear to the students that only on this occasion would the teacher-researcher give the lesson, and that in future lessons it would be with the class maths teacher, and that the researcher's role would be solely and exclusively that of observer.

Once these clarifications had been made, we began an explanation of the software. Some students said they had already used it in the fifth grade, others had used the software but no longer remembered how to use it, and there were still students (new to the school) who had never seen the software. We started by explaining how to open the programme, showing the interface offered by the software, how to use the buttons and drawing pictures on the screen. The constructions were diversified until we obtained more complex figures, or those full of detail, such as star polygons, and we used the animation feature, emphasising that the software allowed us to "drag" the figures, moving them around and also animating them as they were happening on the screen.

After these clarifications, we headed to the computer lab so that the students could explore the software in any way they wanted. Most of them said they also wanted to make drawings and animations similar to the ones they had just seen. The students opened the programme, some with our help, and the class experimented freely with Cabri, making a wide variety of constructions. This time in the lab lasted twenty-five minutes, until the lesson ended.

5.5.1 The observation sessions

As described above, the observations took place in the computer lab, lasting one lesson (fifty minutes). The lessons were designed to cover topics related to plane geometry, with triangles and quadrilaterals. To this end, a didactic sequence[21] was drawn up to develop activities using the Cabri-Geomètre tool. These moments were observed by the researcher, who followed the processes, the dynamics that took place between teacher, student and geometric knowledge, without interfering or helping the students with their doubts. Most of the students in this grade were already familiar with her, as they had worked together in a maths workshop with some of the students in the class the previous year, and because of the meetings at the school's activities, she was not a total "stranger" to the students.

At each observation session, the students arrived from the classroom and were already in the computer lab, where the recorder was being prepared to be placed next to the teacher. We then waited for the class to settle down and for the teacher to start the activities. From then on, we would observe, trying to keep a certain distance from the teacher and the student(s) who requested it at the moment, trying to listen to the dialogues, observe the student's posture in front of the computer and also the interactions between the students. The tape recorder used to record what the research subjects were saying was kept in the pocket of the teacher's coat, in an attempt to ensure that the dialogue was recorded.

The observations were recorded by the researcher in a "field notebook". Photographs were also taken during the laboratory sessions. Some of the activities carried out by the students were

[21] A didactic sequence is a set of activities around a theme and is not limited to a single lesson. In this sense, the didactic sequence comprised 6 lessons.

collected, although the focus of the subsequent analysis was on the relationships and interactions, as it was considered important to have access to the records and work carried out during the interactions, which could provide support and help in understanding them.

By analysing the dialogues (which have been transcribed), it will be possible to detect and identify the relationships that shape the contract established between teacher and students during the sessions using Cabri-Géomètre.

The observation of the sessions, the transcription of the speeches and the "field notebook", in which the events considered pertinent to the context were noted down, are the main elements for this analysis. In addition to the material collected, we sought Guy Brousseau's theory as a link to support the analyses and perceptions, in order to be able to abstract important elements for the research from the phenomena that occurred.

CHAPTER 6

UNDERSTANDING TEACHING CONTRACTS

In this chapter, we will present our thoughts on the episodes chosen from the dialogues and interactions that took place during the observation sessions in the computer lab. The moments of observation and transcription gave visibility to the students' reactions, whether of joy, discouragement or indignation at their teammate's work. As this is basically a study carried out in everyday school life, the data was analysed on a qualitative basis, by describing the information collected. In this way, the results presented tend towards an analysis of the students' behaviour when using the computer lab and do not seek to represent a statistical study. Nor did we seek to specifically evaluate any software or classify it in terms of its performance. The driving force behind the research consisted of investigating the interactions that took place in a computer-mediated learning environment, supported by the Cabri-Geomètre software. The characteristics, particularities and aspects of these interactions will be analysed in the light of the theory of didactic situations.

This analysis will focus primarily on the interactions that took place in the relationships established between teacher and student and geometric knowledge, which, according to Brousseau (1986), is a motivating agent for teaching and learning. The interactions took place against the backdrop of the use of Cabri software, which offered possibilities for working with geometry content in a dynamic way. In this context, the phenomena will be analysed from the point of view of the didactic contract established during the lessons observed. To this end, we emphasise that considerations will be made, describing the phases of a didactic situation, without losing sight of the specificity and totality of the manifestations of understanding geometric content mediated by the computer.

In order to organise the data, categories of analysis were defined, based on the "backbone" of the interactions that took place. We believe that these categories contain elements that were chosen as the focus of this investigation. They are: use of the software and use of the computer; student; teacher; geometric knowledge. The first two categories (use of the software and use of the computer) seem to be redundant, but we would like to clarify that some impasses occurred due to the use of the machine ("the microphone crashed", the screen suddenly closed, etc.) while other moments were characterised by the use of the interface offered by the software. Thus, categories were taken into account which, because they were different, had different implications.

It is worth remembering that these categories are interwoven, hence the need to emphasise certain speeches, which reveal a complexity, in order to understand the didactic contract present there.

6.1 THE USE OF COMPUTERS AND CABRI-GÉOMÈTRE SOFTWARE

As mentioned above when describing the subjects and the research universe, the students use computers at home for various purposes, including schoolwork, and routinely use the Eureka virtual environment. We emphasise this fact so that there is greater understanding when there are references to the skills and familiarity that the students have in front of the computer.

Before the observation sessions, there was a moment (one lesson) when the teacher-researcher explained to the students the reason for our presence in the next lessons. The class teacher preferred us to show the use of Cabri as a form of presentation. We used a multimedia room to present Cabri. Most of the students had already used this tool in the previous grade, but we thought it would be appropriate to carry out a levelling activity in relation to the use of the software. The students were interested and some commented that they felt important for taking part in scientific research. All the students were interested and immediately asked when they would be using Cabri. The answer was that it would be their turn to use the software in the next few minutes.

A few minutes after the explanation, we went to the computer lab. There, some students sat alone at a computer and others were in pairs. As soon as everyone was settled, they were asked to open the Eureka environment, in the Maths subject, in a specific folder containing a script of activities for exploring the software. The students opened the folder and began their activities without any major difficulties. The class teacher followed the whole familiarisation process and helped the students with any questions they had. They all used the tool with great enthusiasm, constructing points, lines, parallel lines, perpendicular lines, triangles, in other words, basic elements. They also worked on constructing and moving some geometric figures, giving them a first taste of the software. At the end of the activity, the students were set a challenge: to draw a picture of some kind, including circles and squares. They were not asked to animate the composition, but many students were keen to use this resource, as the great novelty was movement.

With this first contact, it was possible to see that the students constructed several drawings quickly, being able to easily undo, erase and redo them. They didn't show any fear of making mistakes, because unlike drawing with a pencil on paper, they could quickly redo the lines. We also noticed that the students were very curious to follow their classmates' work, calling them over to see theirs and comparing the tasks to make sure they weren't too different.

This comparison served as the basis for a discussion such as: my square is ABCD and your square is ADCB. The students would then compare the figures and try to find differences and/or points in common. If they couldn't agree, the teacher was asked to clarify any doubts. When a student didn't know how to use a button or the toolbar, another would readily help, promoting interaction and collaboration between them.

At the first meeting, the students left feeling very satisfied and excited. Many said they would rather be playing with Cabri than having to return to the classroom. In this exploratory session, it was possible to observe some aspects expressed by the students:

1. Student acceptability of the interface offered by the software;
2. Motivation towards the proposed activities;
3. Understanding and construction of the initial concepts of Geometry (point, line and plane);
4. Progress in understanding the concepts, as a prerequisite for the following activities.

At the beginning of each lesson, the teacher guided the students through the procedure of opening the programme, until they reached the Cabri screen, and these instructions were repeated until everyone had started using the computer and opened the programme. The following dialogues reveal the students' attitude towards the machine, revealing a didactic contract in which what was expected, implicitly, were work guidelines for the students to carry out. It is from the perspective of the use of the machine that we will first analyse the phenomena relating to the use of the computer and the software at the start of the activity.

With the certainty that the didactic contract delimits and conditions the roles played by the teacher and students in the didactic relationship, we know that it is the teacher's role to create the conditions for students to appropriate new knowledge. We agree with Brousseau (1996, p.49) when he states that it is the teacher's competence to possess and/or develop skills in: "proposing a learning situation to the student so that they elaborate their knowledge as a personal response to a question and make it work or modify it as a response to the demands of the environment and not to the wishes of the teacher". Although, as Brousseau (1996) categorically states, every didactic situation contains something of the teacher's intention and desire, the ideal and necessary thing is for the teacher to get the student to forget the didactic presuppositions of the situation.

At the beginning of each lesson, the teacher explained the aim of the activities and let the students get started. The tasks were designed to encourage the construction of quadrilaterals, with the intention, at each specific moment, of exploring certain elements of polygons, such as angles, sides and vertices, as well as their nomenclature. The time it took to open the programme and start the work was longer than expected, as some computers didn't switch on at the time and the students had to move, as can be seen in the comments below from the 1st observation session[a] . To record the dialogues, the subjects were coded as follows:

Teacher - Profa

Students - Student 1, 2, 3,..., totalling 37 students. As each student takes part in more than one dialogue, the numbers for each student exceed 37, which is the total for the class.

Profa: - So let's go to Start, programmes, programmes,

Student: - *Profª , please!*

Profa :- *Start, Cabri II programmes.*

Student1*:* Profa, Profa, *I didn't open it!*

Profa*:* - *Come on, come here.*

Student2:- *Saved now*, Profa?

Profa: - *Hail!*

Profa*:* - *Come here, come here and tell me what you've done. Tell me, tell me what you've done.*

Student1*: - Start, accessory, Cabri II.*

Profa*: - Start, accessory, what about Cabri II?*

Student1: - I've opened it, it's here, but the machine is having problems. (The machine isn't working)

Prof[1] *: So let's swap* [places]

Student4 to student3*:- You have to save every step!*

In this environment, there were no desks but rows of computers, as can be seen in the picture below:

Figure 8 - Students working in the computer lab with Cabri. .
Source: The author

As mentioned above, the students were familiar with the computer. This familiarity can be seen the moment they enter the computer lab: they arrive, sit down and type very quickly, opening the necessary screens. However, from the above, we can see that the students are dependent on the teacher's guidance. We are faced with a paradox, because from one minute to the next, the student stops mastering the computer and starts being mastered by it, or rather, admits their situation as a learner, and assumes their role in the didactic contract, where it is acceptable for the student not to know. This exchange of roles takes place at the exact moment when the student opens the Cabri screen and is faced with the new, challenging element. At that moment, the students start calling for the teacher.

She turned to them, trying to help them. The first doubts were not about the proposed activities, but about how to open the programme.

According to the transcripts above, the teacher was showing the students how to find and open the programme. Student 5 (dialogue below) was unable to turn on the computer; the teacher suggested that he change places, given that there were 37 students and 26 machines. This student had to move to another seat and join an already formed pair. It can be seen that at this point there is a didactic contract at work, in which the student, without being able to use the computer, already knew that he would have to sit with other classmates and that there he would have to share the tasks of handling the mouse and reading the activities script, which would require him to adapt to the tasks to be completed.

We emphasise that teaching contracts are influenced by the contexts in which they are established, because according to Schubauer-Leoni (1986) apud Silva (1996, p.11) "each pedagogical context determines a specific contract, in correspondence with the characteristics of the school class and the institutional variables that condition its functioning". This statement suggests that in computer-mediated learning environments, didactic contracts are governed by different rules.

The teacher can't solve the "operational" problems with the computer when it won't open, or when it can't be switched on, or even when the screen suddenly goes black; but her attitude is to manage this situation, in other words, to play her role within the didactic contract, which is to ensure that the students are able to carry out the activities using a computer, with the aim of favouring moments that are conducive to learning with dynamic geometry software. The following dialogue shows how the teacher manages this situation calmly.

Student5: *Prof[a] , I'm opening it here and it's just giving me an error on the page!!!*
Prof *Then get out and start all over again!Close and start all over again.*
Student 5*: But I've done it three times!*
Prof *Close, close, close every window, don't minimise, close everything and start again! So...*

content... 4th *bimester, did you click on it?*

Student 5*: I clicked*

Prof[3] *: Now it'll work!*

Student5*: Yes!*

It should be noted that not all moments of impasse when the computer is working are resolved promptly, such as the situation in the dialogue above. When an impasse occurs with the computer and there is nothing to be done immediately, the teacher can use two strategies: put the student to work with another colleague or call the systems analyst to see what is happening. In all the impasses that occurred, the teacher either solved the problem together with the student or asked the students to change places. Although the students tried to solve an operational problem with the teacher because the computer wasn't working, they accepted the fact that the teacher couldn't solve it. A didactic contract is revealed in this context, when teacher and pupils, faced with an obstacle in relation to a technological tool, take on their roles and submit to their obligations: pupils handle the machine showing a certain skill, which may even be greater than the teacher's in this respect. When a difficulty arises, they try to solve it together, but the student understands that it's not the teacher's job to respond to every impasse of this nature, and accepts the need to call on the help of a professional who is better qualified for this purpose, without, however, failing to consider the legitimacy of the teacher's role. However, when an impasse arises regarding the use of the Cabri software, there is a much greater expectation, not to say total expectation, on the teacher, since in the context of using the software there are other rules in force in the established didactic contract, whose norms, implicitly or not, indicate that the teacher should know more about using the software and guarantee information about the geometric concepts involved in the activities.

At this point, the teacher enters "the game" to help the student and provides support for carrying out the activities. This rule of the didactic contract seems to be very clear to the students.

The use of resources in the classroom, such as a computer tool to complement and support the teaching and learning process, implies changes that can alter and/or affect the rules of the didactic contract that has been established. However, according to the observations made, Cabri was used in a supportive way, where the emphasis was on exploring geometric entities.

The main theme of all the activities was "quadrilaterals" and the teacher designed activities to explore the concepts embedded in this theme, with the aim of favouring learning through construction and without attaching too much importance to the use of software.

We would also like to point out that, however modern the software may be and however many advantages it may offer in terms of gain, in terms of the dynamism, movement and interactivity of Cabri, there are impasses that can delay and even compromise the progress of the activity, such as what happened in the first dialogue above, in which the technical problem was not solved and the solution was to move the student. This analysis shows that there are not only advantages to using a technological resource, nor could we position ourselves in a naive way, wanting to present only the benefits of this tool. The obstacles reported also led to frustration on the part of the student, as he had to continue the work in a different way to what he had started, which was a demotivating factor for this student. There is no way for the teacher to foresee this kind of situation, as there is no way of keeping track of each student when they open the programme, given that the dynamics in the computer lab are intense, requiring the teacher to constantly move between the rows, answering the students' calls. However, in this case, the teacher must be vigilant so that the student who was just watching is given priority in the next lesson when it comes to handling the mouse. The teacher's watchful eye is the guarantee that the activities will continue to be carried out in a balanced way, in other words, ensuring that the next time this student gets a chance to actively participate in using the mouse.

The following dialogues show an unforeseen situation that the students were trying to get around. When they couldn't, they asked the teacher for help.

Student 34: Proí'[1] , my *programme has closed!!!*

Prof[1] *: It's not closed, let's see!!!*

Student 34: Proí', *when I was trying to delete the thing, we pressed the wrong key and everything closed!!!*

Prof': *Try to go back!*

Student 34: Prof', Prof', *it didn't work!*

Prof': *Had you already saved it?*

Student 34*: Yes!*

Prof': *Where did you save it?*

Students: *We see!*

Student36 to student35: *That's not it, I saved it under my name.*

Prof': *Look at the previous one, and see where you left off!!!*

The students said they had saved the work, but they didn't know where. The instruction was for the students to save it using Cabri, in a folder for grade 6 students[a] , with the student's name, or by combining the names of a pair. From what she said, the students had trouble finding their work. In the last speech, the teacher suggested that they continue from the previous stage, meaning that, given the time it would take them to locate the activity, they could redo it. These situations, including trying to locate the work, show that the students didn't intend to redo the activity, and if they didn't find it, they would have to go back and do it again.

If a situation like this were analysed in an activity carried out using paper and geometric construction tools, it wouldn't happen and there wouldn't be this frustration. On the other hand, when using software, it's dynamic and easy to erase a wrong construction, in this case much quicker than if it were done with pencil and paper. Once again we are faced with an unexpected situation.

In the observations, we noticed the need for the students to ask the teacher, at every step of the activity, when and how they should save. There is a paradox here, as there are times when the students transgress the contract by being in a situation of mastering the use of the micro as much as, or even more than, the teacher. While there is a culturally, socially and historically established condition that it is the teacher who knows much more than the students, in the situation being analysed, there is complete independence on the part of the student in relation to the teacher's role. It has to be said that this "greater knowledge" on the part of the student is more evident when they look for shortcuts - when opening screens, accessing the school portal where there is a link to the Eureka virtual environment, when switching on the computer to surf the internet - without assuming the role of being a learner all the time. All the students had complete autonomy in dealing with the machine, until they opened the software. At that point, the impression was that the students "remembered" that their role, albeit implicitly, was that they knew little or nothing and that the teacher should help. This change in student attitude, from active to passive, happened with more than 30 students, in other words, the vast majority. As soon as the Cabri screen was opened, the students implicitly resumed their role in the didactic contract, requesting guidance from the teacher.

6.2 THE TEACHER

In order to carry out the activities using the computer, the teacher tried to use the Cabri software at a stage in the learning process when the students already had some knowledge of the subject. The task set for the students was to go to the computer lab to make inferences about the topic they had already seen in class, with a view to reviewing concepts in order to move forward with the activity.

Initially, the teacher used a template for the activity script. This script was divided into stages, with instructions on the procedures to be carried out by the students to finalise each stage. The students worked with two screens at the same time: the Cabri screen and the screen containing the activity script, which consisted of a Word document containing questions and some definitions. After opening the programme and the Cabri screen, the students began the activity.

In the first lesson, a general doubt arose among the students. When they saw the image of the Cabri screen in Word (the teacher copied and pasted it to illustrate how the picture would look), fifteen of them mistook it for the Cabri screen itself, thinking that they could use the software through that image. Based on this doubt, they asked for a more detailed explanation. The teacher was surprised by this, as she hadn't thought of this hypothesis. At this point, the researcher was asked for help and she promptly clarified. So the students continued with the activities. We can even take advantage of this moment in the research to glimpse a clause in the didactic contract. This moment occurred when the students were asked to clarify whether or not the Word screen was the Cabri screen. They all fell silent at the same time in order to listen to the researcher, in contrast to the dynamics of this environment, where the students and the teacher talk, exchanging ideas all the time.

Also, a contract established at this time was that the researcher teacher could give her opinion and that her opinion should be heard, without, however, disabling the role of the class teacher. It was socially and pedagogically acceptable for the researcher to give her opinion on the software, without causing any embarrassment. The roles played were in line with the contract that had been established at the time the observations began. This showed how seriously the students took the intervention, attributing authority to the research and also to the researcher. The students knew that
were the subjects of the research and at that moment they seemed to be happy to take part in the proccss.

This was the only intervention during the observations, because the impartiality that must permeate the moment was known, and the observer's presence went unnoticed, given the non-participant status. Other impasses were managed by the teacher, for example: one student was sitting alone while three others were grouped together to use a microphone. She intervened and separated the trio, suggesting that one of the students go to another computer with the colleague who was alone. In the first two lessons, the activity was guided by the same script, but in the third lesson, the teacher changed the strategy. This was a breach of the rules of the contract, in which she adopted a new strategy, as can be seen in the following dialogue:

Prof[1] : Good *morning sixth grade, schh .... good morning, sixth grade, ..please!!Each computer should open Cabri II. Yes! Today we're going to work with the Cartesian plane. What are we going to work on? Look!*

Student 5*: Amanda, are you going to sit here or next to me?*

Student. 6: Prof[1] *, I'm opening it.*

Prof[1] *: -Sch... Come here, both of you: Afonso and André, come here, both of you...Just a little bit!..Come here, both of you.* [The teacher takes the two students out of the room]

[Teacher returns to the room] *So you see, in Cabri, we have to make... Daniel, Ana, then you can't do it!!! We're going to make the Cartesian plane, which would be what?*

Student 9*: two lines like this. Ó!*

Profl: (on the board*): So there would be two perpendicular lines, right, and within these perpendicular lines, with measurements, that you, that in reality there is no standard, exact measurement, remember that we usually use the centimetre, don't we??as we usually use on graph paper; what are we going to do???*

After the perpendicular lines, we're going to mark the x-axis, y-axis, in other words, the abscissa axis and the ordinate axis, and then we're going to graduate this Cartesian plane, graduate it, and within these planes I'm going to assemble it! Gabriel, are you paying attention?

Student. 10*: Yes*

Profl*: OK then! And what am I going to have to do within these planes? I'm going to make geometric shapes, so what am I going to make: a square, a rectangle*, exactly the same as the previous lesson we had, only now all the figures have to have the Cartesian plane at the bottom. In our previous Cabri, the teacher would *give you hints and you would follow the procedures, but* <u>*now you don't have to*</u>. *The pair will use the measurement of 1 cm, 2cm, 3cm, 4cm, whatever they think is best, then they will draw here: the square, the rectangle, the parallelogram, what else here*?

Student. 12*: rhombus*

ProfP-That*'s right, and what else?*

Student13. *-Trapezium*

Proí'[1] .*-Trapezium, so look: you've drawn it, and what are you going to do? Colour it in!*

Student 14*: Wow, that's great!*

Prof[1] *: You can either colour around it, because you can work with the coloured segments, or you can colour inside it. Don't forget that for each geometric shape, what will happen? You'll have to mark points A, B, C and D, in other words, the vertices of the polygon. Any questions?*

Student 14*: No!*

Prof[1] *: No?Then good lesson!It's exactly half past seven, and you have half an hour to do the activity!Don't forget to save!!!*

This dialogue between the teacher and her students describes a change in the methodology used to carry out the previously proposed activity, showing a rupture in the didactic contract, because we noticed that the teacher realised that in the previous lessons the procedure had been very "directed", and now, after two lessons, she observed that the students were fully capable of walking on their own, a situation that required a change in the teacher's actions.

We looked to Pinto (2003) for a better explanation of the new stance adopted by the teacher and the implications of this fact, since the author draws attention to the richness of these situations which involve breaking some of the rules of the contract, with a view to the progress of the students' learning, after some changes resulting from the teacher's attentive look at the students' cognitive characteristics. "The function of a contract is to manage these relationships, not by plastering them, but by making them progress, putting them in tension, through a series of ruptures" (PINTO, 2003, p.101). We believe that the teacher played a fundamental role in relation to the didactic contract in force at that time in the classroom, altering her teaching action and demonstrating that she was open to change.

Kenski (2007) points to the need for a change in practice when carrying out activities involving technology. The dynamics surrounding the use of technological resources can be different, but they can also be lost by not having a clear and appropriate objective from a pedagogical point of view. The author states:

> In order for information and communication technologies (ICTs) to make changes to the educational process, they need to be understood and incorporated pedagogically. This means that it is necessary to respect the specificities of teaching and of the technology itself in order to ensure that its use really makes a difference (KENSKI, 2007, p.46).

Still referring to the dialogues above, we noticed that the teacher, in wishing the class a good lesson, seemed to be sending out a message such as: work on your own, show that you can do it! We also notice in this dialogue the paradox of an action situation, because the student is now faced with an interesting work proposal when he says "Wow, that's great!" (student 14), and also a challenging one, which must be done at the moment, triggering an action situation; however, in order for this action to take place, it will most probably be necessary (and the dialogues confirm this) for the teacher to help, to intervene. As mentioned above, in the action phase, the student is faced with a situation in which, in order to solve their problem, they need to know what they want to teach. In such situations, the experiential, argumentative aspect prevails, while the theoretical aspect of the concepts involved remains at a distance. Initially, the teacher points out to the students that this topic (Cartesian axis) has already been worked on in class, but in a different way, with a different approach, and that now the students have to work on combining two topics: quadrilaterals and the Cartesian plane.

The paradox occurs when, by putting the students in an action situation and leading them to work, the teacher is implicitly aware that they will question her several times and she will help them. This paradox is experienced only by her, who gives her students the opportunity to carry out an activity independently, but prepares herself for the demands that will come. As we can see, the activity was prepared in such a way as to require the students to reflect on the construction, using the concepts and creating resolution strategies. The perception of this fact is based on Brousseau's words:

> The teacher's job, then, is to offer the student a learning situation so that they can develop their knowledge as a personal response to a question, and make it work or modify it in response to the demands of the environment and not to the teacher's wishes (BROUSSEAU, 1996, p. 65).

The teacher needs to get the students to forget the didactic assumptions of the situation and see the work as their own. This activity allowed the students to experience the return stage of the situation, when they put a lot of effort and care into their constructions, even leaving their seats to look at the geometric shapes obtained by their classmates and compare them with their own, a fact that can be seen below:

Student12: Prof, we*'re doing this challenge here, come and see!*
Student.13 to student.12*: This one is flashing, how did you do it?*
Prof[1] *: Look how cool it is, it's flashing! That's a star polygon!*

We emphasise that devolution is a situation where the student takes on the challenge and the problem, which now becomes their own and is no longer the teacher's desire or intention. Although it sounded like a fun activity, the challenge consisted of drawing quadrilaterals on the Cartesian plane and the students invested in the artistic side, adding colour and movement to their drawing, without, however, neglecting the characteristics of a quadrilateral, which, for example, could not be either a triangle or a pentagon.

Another aspect observed in the dialogue above is when the teacher removes two students from the classroom. They were talking insistently, even after the teacher asked them to stop.

Prof[1] *: - Come here, both of you...Afonso and André, come here, both of you...Just one...*

*(*the teacher took the *two* students out of the room*).*

By removing the two students from the computer lab, the teacher makes it clear that even though they are in a different environment from the classroom, the main objective is learning, and that she will not allow the students to change this by talking too much. Although the focus of this research is on the interactions inherent in the didactic contract, one cannot fail to scrutinise the fact that it refers to the pedagogical contract. The students know that there are rules of behaviour and norms in force in the school. Here again, we can draw on Pinto (2003), who explains the dimensions of the pedagogical contract, based on Filoux's studies:

> Filoux's contract, in showing the existence of the regulation of rules between teacher and student about the teaching project, observes that the pedagogical contract brings implicit power relations whose negotiations, not always made explicit, are already previously established in the institutional contract, which has already defined the status of the teacher and student in relation to their roles in the institution. It is a mutual consent of the rules necessary for the functioning of the school (PINTO, 2003, p. 99).

Most of the time, students are aware of their role within the school and know their obligations and attitudes within the institution; this doesn't mean that they don't want to transgress certain rules of behaviour. In other words, students aren't motivated all the time to comply with certain rules, and we understand that the environment where learning takes place, mediated by the computer, is even more conducive to "forgetting" agreements, rules and attitudes to be adopted. In this case, there was an attempt to use this moment to "break" the rules of the pedagogical contract, but the attentive teacher didn't allow this to happen. She removed the students from the classroom, finished explaining the lesson to the class and then went to meet them again to resume, as the following dialogue shows:

Prof[a] :And *don't forget to save!!! And one thing, I'm going to do exactly what I did with the two over there, oh, they're going to stay outside, if I have any more trouble here in this room, OK?*

Student 7: *Who stayed outside?*

Prof[1] *:All right? One more, one more nuisance I have... You're here to work, not to play games!!!Respect everyone's work!!!Have a good lesson everyone!!!*

(The teacher addresses the students waiting outside)

Prof[1] : Now, *the next time you have to say something to someone, disrupting the lesson, I'll take you to the coordinator.*

Student 3*: But I was just going there to talk to him.*

Prof1: *I don't care, you're talking too much. And...everyone mind your own business if you're not going to be left out.And now back to the classroom and let's get to work!!!*

Students*: Yes...*

(The students quietly go to their computers to work)

In this episode, we saw the necessary intervention by the teacher to maintain order in the class and ensure a favourable environment for the activities to take place. This showed that the didactic contract is permeated with relationship rules that must be complied with and which are already explicit in the pedagogical contract. This situation reveals that it is not possible to analyse the didactic contract without considering the particularities of the pedagogical contract previously defined by the institution.

Analysing the teacher's subsequent conversation with the two students, we noticed that she made it clear that it would be one more chance, and if the students didn't behave appropriately, they would be referred to the coordinator. We also noticed that after this intervention, the students entered the classroom embarrassed and began their activities, and the class was quieter, looking at their classmates with a look that said: "Now, because of you two, we're going to have to be quiet and the teacher is angry." The teacher's tone of voice remained serious for a while longer and, little by little, the class began to move again, asking questions of the teacher and their classmates.

Here we present another moment that reveals another breach of the didactic contract, experienced by the teacher and also accepted by the students. The following dialogues include a situation in which the students were at the start of the activity that the teacher had proposed, which consisted of constructing a Cartesian coordinate axis in order to build quadrilaterals on it. The activity was starting, the students were doing the activity when a student called the teacher over and revealed a "discovery" to her:

Student 5: *Prof*[1] *:, look here on this button!!!*
Prof[3] *: What's that?*
Prof: So what?
Student5*: Here, oh, if I click here on show axes then it gives the graduated axis!*
Prof[3] *: (Smiling) Cool!!!*
Prof[3] *: Personaaaaal look here, show axes, in the last icon, it already shows the whole graduated axis!!!*
Students: *Wow! Ooooohhhhh! What a ten!!!!*
(a few dialogues later)
Prof3*: Do you know what else you can do?*
Student7*: No*
Prof3: *Go to the last button, and show axes! And it's already graded, see?*
Student7*: Now?*
Prof[1] *: Yes, go to labels! Click on the last one to see it!*
Student7*: Shall I click on the last one?*
Prof[1] *: Yes, go on!*
Prof[1] *: Click on the last one, show axes and scroll down.*
Student 7*: Here?*
Pupil9 to pupil8*: When you get to the last one, the axle is ready!!!*
Profl: *Go to yours, to show axes and it's already graduated, and you don't have to divide it up!*

One student realised that there was a "show axes" option, which generated the complete axis as well as the graded axis; realising how easy it was to get the axis, he called the teacher, who couldn't contain her surprise. She immediately socialised this discovery, and from the surprised tone in her voice, she made it clear that it was the student who had found an easier way to conduct the activity, since the focus of the task was on constructing quadrilaterals, not the Cartesian axis. The rupture in the rules of the contract that had been in place until then occurred when, even without stating it, everyone knew about the new strategy she had adopted, and that the change in direction came from an unexpected situation. She could have kept the initial proposal, which was even more labour-intensive, but no, she admitted that it would be better for the progress of the activity and, from that moment on, she adopted the new strategy and shared it with the class. The student smiled, as if to say: "*I managed to contribute a lot in this lesson, to the point where the teacher made a substitution in the activity proposal*".

The contract that permeated this episode was broken, since its rules were broken to make way for others that, in the teacher's judgement, contributed to a better flow of activities. This can be explained theoretically by Brousseau when he explains the clauses for breaking a didactic contract:

> In particular, the clauses of rupture and the framework of the contract cannot be described in advance. Knowledge will be precisely what resolves the crises resulting from these ruptures that cannot be predefined (Brousseau, 1996, p.52).

Again, according to Brousseau, each piece of knowledge must correspond to specific situations (problems) and probably didactic contracts that "evolve, break down and change at the expense of covering learning situations while maintaining the coherence of actions with the didactic assumptions involved in the intention of promoting learning" (p.65).

6.3 DIDACTIC SITUATION

As already mentioned, Brousseau (1996) makes an analogy between a teaching situation and a game, in which students and the teacher play, and there may also be a game between the students. In a closer relationship with knowledge, "the game must be such that the knowledge appears in the form chosen, as the solution, or as the means of establishing the optimum strategy", which refers to certain questions, but above all, according to the author, "the specific game of knowledge must justify its use or its appearance, in accordance with theoretical didactics." This statement is an alert to the seriousness with which any learning situation must be treated and also the vigilance with which the teacher must develop; firstly in relation to their epistemology, their knowledge that will have a direct impact on the activity/lesson to be proposed. It is from this perspective, looking at a learning situation, that the dialogues deemed pertinent to understanding the rules that permeated the didactic contract will be analysed.

What follows now, according to the speeches, is a formulation/validation situation led by the teacher. In the first speech, there is a mixture of two situations: by referring to triangles, the teacher is trying to validate learning that has not yet been institutionalised. She is referring to the sum of the internal angles of a triangle, which the students have already constructed and validated, but she has not yet generalised what the students have perceived in relation to triangles, using this acquired knowledge to serve as a parameter for formulating a new topic, which is the sum of the internal angles of a quadrilateral. This is shown in the dialogue below:

Student 17: Let's see if we can do it now!

Prof[1] [some time later]: *Sixth grade, attention... now that you've constructed the four types of triangles and checked what the sum of their internal angles is, we're going to check what the sum of the internal angles of a quadrilateral will be like. Will it be the same? Will it be greater? Will it be less than 180°? I want you to finish the triangles and start working with*

Draw the quadrilateral and define for me what the sum of the internal angles of a quadrilateral looks like. Firstly, what is a quadrilateral?

Students: a square

Prof[1] : it's a..

Students: squares

Prof[1] *: Is it only squares?*

Student *18: All the ones with four sides!*

Prof": *Very good, Mr Felipe, but when we talk about quadrilateral, it's only square, only square, only square! Poor rectangle! Poor parallelogram! Poor others!*

When referring to the phrase "Is it going to be the same?", the teacher takes 180° as a reference, obtained in relation to the triangle, leaving a doubt in the air, or rather, a proposition that from now on needs to go through situations of action, formulation in order to be validated and finally institutionalised.Still in the dialogue above, we observe that the teacher tries to show the students the variety of quadrilaterals, not just the square, and she does it in a playful way.

In the following dialogue, we have a *situation of institutionalisation*, promoted by the teacher.

Prof[3] *: But that's a quadrilateral...*

Student *22: Isn't that the parallelogram?*

Prof3*: And the parallelogram has two by two parallels; the parallelogram is as if you took a rectangle and squeezed it, heim!!!!!Sch!!!Peopleaaaal, attention here, for those in doubt, let's clarify! Attention João Vítor, that's enough! Felipe, throw that chewing gum in the bin, please now! Come on! From the square, we're going to transform it into...what polygon is this?*

Student23*: A square.*

Prof3: *Come on, people!*

Prof3:(*on the board): We have a square and we're going to turn it into a rhombus! Look what happens! What have I done? I'm going to move the three centimetres from here to here. Look what happens!*

Student24*: It stretched one side and shrunk the other!*

Profano quadro]:What I *did; from 90° the angle went to acute and obtuse.It's as if I had a square and I had squeezed it, OK! Now look...you have to construct a rectangle, it's here.sch...I'm still listening to someone talking...and...Bruno...look at this rectangle and then he asks you to make a parallelogram, so what does he do?*

asks: validate this construction; in what way? How am I going to show that the two polygons have the same area? I'm going to bring, I'm going to move the parallelogram on top of the rectangle. Who has done this?

Student 25*: I did it!*

Prof[1] *: Heloísa has already done it.*

Prof[1] *(on the board): The area of the rectangle will be the same as that of the parallelogram, why? Because this rectangle here is going to be exactly like this... on top of this one. That's why it asks: overlap, what is overlapping? The exercise asks us to do this so that when we place one figure on top of the other, we can realise that the area is the same!*

The main aim of the activity was to explore the areas of quadrilaterals, but first students had to construct the figure. The students confused trapezoids and parallelograms, not associating the name with their properties. The teacher called the class together and recapped the concept of a parallelogram, but in very simple language, emphasising the main characteristics, which are parallel sides and acute and obtuse angles. What's more, she started from knowledge that most people already had, which was to take the square as a reference and then "deform" it, transforming it into the quadrilateral in question. Her intervention constitutes a situation of *institutionalisation*, in which the teacher formalises knowledge, makes it official and ready to be used conventionally. Brousseau (1986) reinforces the importance of the teacher institutionalising, and can even do so at different stages of the didactic situation:

> Institutionalisation takes place both in a situation of action - the value of a procedure is recognised and it becomes a reference resource - and also in a situation of formulation. There are formulations that will be preserved ("this is how it is said", "that must be remembered") (BROUSSEAU, 1996, p.56).

We would also add that in situations designed to promote learning, the phases of a didactic situation (action, formulation, validation and institutionalisation) are often intertwined to the point where it is not possible to establish their boundaries, i.e. where one ends and another begins.

Again in the dialogue above, we can see that in relation to areas, the teacher is in a formulation situation, and by overlapping figures (action situation) she provides the conditions for students to deduce that the area of a rectangle is the same as that of a parallelogram.

6.4 ERROR HANDLING

One aspect observed and considered important to address in this work concerns the errors made by the students, whether as a result of using the software or because of the gap in geometric terms. Referring back to the theory of didactic situations, errors permeate both didactic and a-didactic situations, because errors and mistaken conjectures can occur in situations of action, formulation and validation. However, in the situation of institutionalisation there is no room for error, because it is at this moment that the teacher confers a status on the knowledge, legitimising it and giving it a universal character. This is the opportune moment to correct possible distortions (misconceptions, incorrect construction) that occurred in the previous phases.

The following are dialogues in which a wrong move was made by the student, which cannot be attributed to his handling of the mouse buttons, but rather to the strategy he adopted to solve the proposed problem situation.

Student 15: Back again, to the picture?

Prof[1] : No, *go back and read it. Do they overlap in the picture?*

Student 15*: Hmm...*

Prof[1] *: That's right, OK, go on.*

Prof[1] *: Hey, this is wrong, you know? It's because for it to have the same area, this one should be here and this one should be right on top. See?*

Student16 to student15*: See, I didn't tell you it wasn't right!*

Prof1*: So my love, this little dot on top of this little dot and this little red one on top of the little red one here.*

Student 16*: Ahaan...*

Prof1*: That's right, now it's right... now lift it up and now put it down a little bit. Now it's right, that in reality if you take this little piece of the parallelogram, and pass it over here, it will turn into a rectangle, you see? That's right, now it's right. Don't forget to save it!*

Student 16*: Yes.*

Prof1: *Yes! Look, see? Isn't the area the same?*

Student16: *Yes!*

Prof1: *Right!!! And don't forget to save it. A square is actually a rectangle.*

The activity involved overlapping figures in order to explore their areas, and the student didn't realise that they weren't overlapping yet, and his classmate retorted, as if warning him that it wasn't right, but even so, the classmate needed the teacher to validate his action, or to say that it wasn't right. This reveals a type of didactic contract whose rules are paradoxical to a-didactic situations. Firstly, the students were in an action situation and formulating their ideas around it, without the presence of the teacher, characterising an a-didactic situation, when an impasse arose: the students disagreed about a certain step in the activity, so the teacher was called in to clarify the problem that had arisen.

The student's dependence on the teacher's approval is part of this process, but the paradox is that the students, as much as they know how to use the software and are working in a way that shows relative independence, all it takes is one impasse to "break" this routine, and there is only one way out: to call the teacher, because it is implicit in this didactic contract that the problem that challenges the students contains elements that are known and others that are unknown and require the teacher's help. This is due to the way in which problems, in general and at different levels, are designed; problems that ask for simple answers, without deeper thought, or without the need to look for strategies to solve them.

In these statements, it was possible to realise the mobile nature that Cabri, a dynamic geometry software, offers. Most of the students, without realising that by moving the figure or dragging it in order to overlay it on another, could deduce elements necessary for understanding equivalent figures. The activity explored the area of polygons without using the formula, but using the measurements of the sides and the figure area option. The students also, without realising this particular feature of the software, constructed the quadrilaterals by joining line segments, and when they clicked on the figure area option, they realised that the software didn't give the area, but the side measurement. The students, in the action and formulation phase, didn't understand the reason for this disparity. The teacher had to intervene to explain to them that in order to obtain the area, the figure had to be drawn as a regular polygon, not as consecutive line segments. Laborde (1994) deals with this issue clearly, showing us the difference between a drawing and a geometric construction. He states that a drawing, even a geometric one, can be interpreted in various ways and "perception, in particular, interferes in the development of an interpretation when the reader does not have in-depth theoretical knowledge of geometry, which allows them to go beyond the first perspective reading". (LABORDE,1994,p.53) This differentiation happened when the student wanted to calculate the area of the figure and the software couldn't find it; nor could it, because the student's construction was having two interpretations: for the student, it was a quadrilateral, because it was a closed polygonal line, but for the software it was line segments connected consecutively and not the polygon, since the software has the option "regular polygon" in its menu, to avoid this dubious interpretation.

The teacher provoked this situation, because she knew about it, and let the students reach this impasse, until one student realised and validated his discovery with the teacher:

Student 34 to student33: *let's call Prof[a], Prof[a], look here!*

Prof[1] *: Hassan, did you do it any other way?*

Student 34*: I did it here with the rectangle*

Prof[1] *: How did you get square centimetres?*

Student 34*: I went to area and clicked on the polygon!*

Prof[1] *: Bruno, click on the area, then click on one of the sides.*

Student 34: *Prof^a, before clicking on area, you have to go to polygon, because if you go to segment, it doesn't give you the area of a segment!!!*

Prof[1] *: Bruno, let's go to the area; before clicking on area, you have to reinforce it; you have to go to polygon, A, B, C, D, pass over the figure again, so that it gives you the area; you're doing the angles, so it won't give you the area.*

Student 35*: Will I have to do it again?*

Prof[1] *: no! Where are the ones you made?*

Student 35*: here.*

Prof[1] *: so, open it up.That's it..take the values of the angles ;That's it, now reinforce the segments, passing through each vertex, and go back to the beginning AB, BC, CD and DA, reinforce, that's it... but first click on polygon, so that he understands that the area will be of this polygon.Did you click on area?*

Student 35*: I'm going*

Prof[1] *: Now click on the new area.*

In this last line, the teacher makes it clear to Student 33 that she didn't realise this subtlety. In this way, she has given the student a formulation situation, since he is convincing himself that if he used this procedure, he would get the area of the figure; but with Student 34, he is in a situation of validation, because after formulating with Student 33 (who wasn't convinced by what was said), he called the teacher so that she would approve what was elaborated in a situation of action and formulation, since this Student (34) was trying to convince someone of the validity of his arguments, and his colleague (33) didn't understand, and he then believes that the teacher will certainly be convinced.

We would add to the above paragraph a particular feature of the software that directly affects the interactions between the students and also the dynamics of the classroom, as the students encounter certain deadlocks, often generated by the "deformation of the figure" when moving it, which would not occur when drawing with pencil and paper. A priori, it seems more "comfortable" to work in front of the computer than to actually construct the figure with the appropriate tools. However, with each construction with Cabri that is not based on the properties that define it, the drawing does not become a geometric object and is therefore disqualified, deformed, thus generating questions and the need to redo the activity.

This relationship between the drawing and the geometric object has already been made more explicit in this study, but as the subject came up in the dialogues portrayed above, this differentiation was therefore deemed necessary and we also reinforced the false impression one might have that Cabri treats geometry superficially, and that it doesn't require knowledge of geometry to develop activities. The software allows geometry to be explored through the manipulation of basic geometric objects and preserves the geometric properties of the objects that have been constructed. These characteristics, according to Laborde (1994), help students to differentiate between drawing (image reproduction) and figure (image construction) according to the geometric properties that define them.

6.4.1 Situation a - didactic

We should point out that during our observation sessions, the tape recorder was kept in the pocket of the teacher's coat to record her dialogues with the students. While the teacher attended to this or that student, the others worked and the pairs talked about the topic, agreeing or disagreeing, but trying to carry out the activity. In the transcribed conversations, we basically heard the teacher and, occasionally, the students around her. We did, however, get a few lines that suggest that the students were facing an a-didactic situation. We would emphasise, however, that the observations showed that this environment was conducive to the development of situations in which students discussed problems with each other or with other groups. The interaction was intense, with students leaving their computers to go and see another pair's work, to orientate themselves or to disagree. Unfortunately, we weren't able to capture the richness of these moments in any dialogue, but it was possible to see this practice in the observation sessions. We couldn't fail to mention this particular aspect of the phenomenon, which was the interaction that took place between the students without the teacher interfering. The attitude of the students in this environment has already been mentioned, but we feel it is appropriate to emphasise that there was no trace of a traditional classroom in the lessons, from the way the students occupied the physical space of the classroom to the frequency with which they got up, sought help from their classmates or called the teacher. We observed the dexterity with which some of them carried out the activities, as well as the patience and collegiality with which these same students were willing to help classmates who were unable to progress. The teacher allowed these interactions to take place and avoided interrupting unless asked. Let's take a look at the dialogues below:

Student 42 to student 43: What are you doing?

Student 44 to student 42: I'm making a square!

Student 42 to student 43: It's not like that, Nailia, it's not like that!

Student 43 to student 42: Yes, it's a square! Look... the sides are right!

Student 42: *Profª, I need help!!!*

Prof[1] *: So come on, what do you have to do?*

Student 43*: A square!*

Prof[1] *: Then you can start!!!*

Student 43*:And where do I have to go?*

Prof[1] *: Right there!!! In segment...Yes!*

Student 43*: So why didn't it work before?*

Prof: *I didn't see what you did before, but let's go over it again!!!Let's go to segment!!!And now we're going to move this one over here, OK!!!And if you make a mistake, go to edit, undo.Here, on this one, up to here.Then, OK, you go to label, and what do you do???*

The students were discussing the construction of a square; this didactic situation was centred on a drawing, which under geometric analysis did not correspond to the representation of a square; this for Student 42, because for the other, Student 43, the drawing was considered to be the representative of a geometric object, in this case, the square. And her argument was Look, the sides are right! The other student didn't argue back, because what did she mean by the sides being right: straight? The same length? Parallel? The students were faced with an impasse, which was only resolved with the teacher's help. Although mentioned above, Brousseau's definition of an a-didactic situation is repeated here;

> When the student becomes capable of putting into operation and using for himself the knowledge he is building, in a situation not foreseen in any teaching context and also in the absence of any teacher, then what can be called an a-didactic situation is occurring (BROUSSEAU, 1986, apud PAIS, 2002, p.68).

As a-didactic situations are characterised by moments of learning, permeated by situations of action, formulation and validation, it can be said that situations of this type were seen in the observations, and in a very natural way, with a view to carrying out the activity, accompanied by situations of learning.

The analyses carried out so far show that the environment in which the interactions were observed, i.e. geometry lessons with the support of software, includes moments of establishments and ruptures of didactic contracts. The fact that the students are arranged differently from the classroom structure does not inhibit their interaction. This environment also requires the teacher to be dynamic, to move around, because as the dialogues show, she was coming and going between the micros all the time, helping and guiding them through the activities.

The contracts highlighted above point out that the environment can be favourable, but it is up to the teacher to be willing and attentive to notice and act immediately, in order to break with what is not good and allow another type of contract to emerge, with actions from the teacher and reactions from the students, not always expected on both sides.

We therefore affirm that it is the teacher's posture and epistemological vision of teaching and learning that can allow didactic contracts to be broken. If the teacher keeps his or her teaching focused on meaningful learning, he or she will be willing to "strip off" their entrenched and crystallised convictions in order to break with certain attitudes and look for alternatives, tools and support to improve their praxis.

CHAPTER 7

FINAL CONSIDERATIONS

This research aimed to study how the didactic-pedagogical relationships between students, teachers and mathematical knowledge are established in the context of the geometry teaching and learning process, using the Cabri-Géomètre resource.

In order to study the relationships that took place, from the perspective of the didactic contract, the theoretical foundation was based on Guy Brousseau's theory of didactic situations, which provided the theoretical support and starting point for analysing the interactions observed.

In order to gain a theoretical understanding of the themes that permeated the object of the research, geometry, information technology and didactic-pedagogical relations, we used theoretical contributions: regarding the teaching of geometry in Brazil, we searched, among others, for articles and dissertations, such as the work of Pavanello (1989), in the dissertation of Nunes da Silva (2006), Leme da Silva (1997) and Pereira (2001)[22] (. We found contributions from Fainguelernt (1999), Laborde (1994) and Gravina (1998) that approached geometry from the point of view of using software. Fainguelernt (1999) developed a work using geometry with the Logo language, while the last two authors use Cabri and consider it to be a dynamic geometry software capable of offering dynamism and exploring geometric entities in a very effective way.

In order to understand the use of computers in schools, Papert (1994) provided support with his theory involving constructionism, which can elevate the use of computers to a tool capable of helping students to construct knowledge by carrying out an activity that allows them to build something. Valente (1999) helped us understand the arrival of computers in schools and warned that teachers should be careful not to maintain their traditionalist practices and use computers in the belief that they are being "modern". We looked to Kenski (2007) to outline the urgent need for teachers to change their practices in order to use technology.

The fieldwork was carried out using a qualitative approach, with priority given to analysing the observations and speeches of the subjects involved, in this case 37 6th graders[a] and the maths teacher in charge of the class at a private educational institution. The data was collected through observations of six 50-minute sessions, in which the subjects' statements were recorded. The main focus of the observations was to describe the forms of mediation that took place during the computer activities; to identify the roles of the teachers and students during the use of the software; and to

[22] The other dissertations we consulted are listed here in alphabetical order of their respective authors:Alceu Cotta Júnior, Ma Célia Leme da Silva, Evandro A. Bertolucci,Marcos Lúcio de Castro Fonseca,Rúbia B. Amaral Zulatto,Marcelo Carlos Gimenes,Leonor Farcic F.Menk, Maria José A. Souza.

characterise and analyse the didactic contract established between teacher, student and knowledge during the use of the software.

Analysing the data showed the existence of various didactic contracts that were established and broken during the course of the lessons. Breaks in the rules of the didactic contracts came from the teacher, pointing to the importance of the teacher's attitude towards the new challenges brought about by technological resources: not just changes in the way of thinking and acting, but having attitudes that are open to change in the face of new situations arising from the use of computers.

In these moments of breach of contract, the teacher took over the change, and to our surprise, the students continued to confer status on the teacher, i.e. at no point in the breach of contract did the students stop considering the teacher as the conductor of the process, and consider her as the authority in the process. On the contrary, for them, this change meant the possibility of more productive learning.

The study showed that the teacher's attitude is a determining factor in the good (or better) development of activities with the new teaching resources coming from technology. During the observations, situations were noted in which the teacher suggested interactions between students with a view to comparing procedures. This teaching approach seemed to be in line with the constructivist vision, as it encouraged students to build their knowledge through interactions with their peers, providing them with moments of co-operation and exchange of reasoning. As the observations indicate, the teacher's attitude was aimed at ensuring the student's involvement in their learning: by not knowing the immediate answer, she favoured her educational project, opening up space for the student to make conjectures when constructing knowledge.

As far as geometric concepts are concerned, we noticed an evolution throughout our observations, both in the way the students referred to quadrilaterals and in the confidence they showed when constructing or moving the figures. Although we observed a maturing on the part of the students, the activities developed with Cabri brought to light gaps in relation to the basic concepts of geometric entities. The idea that we can move the figure around, build it with some ease, cannot be extended to the basic concepts of geometry. Having some dexterity in handling the mouse and knowledge of the construction menus offered by Cabri does not guarantee that constructions and validations will occur correctly. When using the software, students need to have a grasp of the fundamental elements of geometry, far beyond points, lines and planes. The subtleties and specificities of Cabri are very different from drawing with pencil and paper. To build a square, for example, the student can obtain it using the "regular polygon" option, or through the circles. Otherwise, when the figure is moved, the square falls apart.

These gaps indicate that the software does not mask anything, on the contrary, it reveals the existing gaps in relation to geometry, thus serving as a point of reflection and recovery of those concepts not yet appropriated by the students.

With regard to the students' motivation, the research confirmed what other studies on the use of software have already pointed out. The students are motivated, get into the "game" and feel challenged to solve the proposed problem. The software on its own does not guarantee motivation, but the proposition of a stimulating activity, which leads the student to look for solving strategies, are crucial points if the optimal use of the software is to contribute to the students' involvement in the activities.

The research reaffirms that the successful implementation of technological resources in education depends on three important aspects: the institution, the teacher and the student.

Schools need to include in their political pedagogical project the definition of strategies for the introduction of new technologies and consider them as an appropriate cost-benefit ratio, seeing technologies as an opportunity to boost education, contributing to the formation of autonomous and critical students.

Teachers need to be aware of the importance of their role in the context of educational technology and look for ways of approaching the pedagogical use of technology, improving themselves in this universe, so that they can incorporate the creative use of tools into their practice, in other words, educational technologies, being attentive to new forms of didactic-pedagogical integration, in which teachers and students become partners in the same process of constructing knowledge.

This research was not intended to exhaust the subject. On the contrary, it opens up possibilities for further research that can describe the interactions that take place in a classroom with the aid of a computer. One suggestion would be to investigate the didactic-pedagogical relationships in the teaching of Geometry, from the point of view of the Theory of Didactic Situations in a teacher training course, with Maths undergraduates as students.

REFERENCES

ANDRADE, J. A.A, NACARATO, A M. Didactic-pedagogical trends for the teaching of geometry. In: ANPED ANNUAL MEETING, 27, 2004, Caxambu, MG. Texts of papers and posters. Available at: <http://www.anped.org.br/reunioes/27/gt19/t197.pdf>. Accessed on: 08 October 2007.

AREA,Manoel.Twenty years of institutional policies to incorporate information and communication technologies into the school system.In:Tecnologias para transformar a Educação.Porto Alegre:Artmed,2006.p.153-175.

BERTOLUCI, Antônio Evandro. Teaching and learning geometry: an experience with Cabri-Géomètre software. 2003. Dissertation (Master's in Education) Federal University of São Carlos, São Carlos, 2003.

BRAZIL. Secretariat of Basic Education. National Curriculum Parameters: Maths. Brasília. MEC/SEF, 1998a.

_____ . Parâmetros Curriculares Nacionais: terceiro e quarto ciclos do ensino fundamental: introdução aos parâmetros curriculares nacionais. Brasília, DF: MEC/SEF, 1998b.

BOGDAN, R.; BIKLEN, S. Qualitative Research in Education: an introduction to theory and methods. Portugal: Porto Editora, 1994.

BONGIOVANNI, Vincenzo. Discovering Cabri-Géomètre II: Activities notebook. São Paulo: FTD, 1997.

BROUSSEAU, G. Fundamentals and methods of the didactics of mathematics. In: BRUN, Jean (dir.) Didáctica das Matemáticas. Lisbon: Instituto Piaget, 1996. p. 35-113.

_____ . The different roles of the teacher. In: SAIZ, In; PARRA, Cecília (Org). Didactics of Maths: psycho-pedagogical reflections. Porto Alegre: Artes Médicas, 1996. p. 48-72.

CONTRACT. In: FERREIRA, Aurélio Buarque de Holanda. New Aurélio dictionary. Editora Nova Fronteira. 2. ed. Rio de Janeiro: Nova Fronteira, 1993.

D'AMORE, Bruno. Elements of maths didactics. Translation by Maria Cristina Bonomi. São Paulo: Ed: Livraria da Física, 2007.

FAINGUELERNT, Estela Kaufman. Maths education: representation and construction in geometry. Porto Alegre: Artes Médicas Sul, 1999.

FONSECA, Marcos Lúcio de Castro. The use of computer technology in the classroom: a study of geometry in primary school using interactive learning resources. 2001.Dissertation (Master's in Production Engineering), Federal University of Santa Catarina, Florianópolis, 2001.

GRAVINA,Maria Alice; SANTAROSA,Lucila Maria.A aprendizagem da matemática em ambientes informatizados,1998.IV Congresso RIBIE-1998.Disponível em:http://www.c5.cl/ieinvestiga/actas/ribie98/117htm>.Acesso em 20/11/2007

IGLIORI, S. The notion of epistemological obstacle and maths education.In: MACHADO, S. (Org). Maths education: an introduction. São Paulo: Ed. PUC-SP, 1999. p. 89-113.

KALINKE, Marco Aurélio. Internet in education. Curitiba: Ed. Expoente, 2003.

KENSKI, Vani.Moreira.Educação e Tecnologias.Campinas-SP.Papirus,2007.

LABORDE,Colete.CAPPONI,Bernard.Aprender a ver e a manipular o objeto geométrico além do traçado no Cabri-Géomètre. Revista em Aberto.Brasília,ano4,n62,p.51- 62.abr./jun,1994.

LÚDKE, Menga; ANDRÉ, Marli E.D.A. Pesquisa em educação: abordagens qualitativas.São Paulo: EPU, 1986.

MANUAL do Cabri: o caderno interativo para ensinar e aprender Geometria. São Paulo: Ed. PUC, 1991.

OBSTACLE. In: FERREIRA, Aurélio Buarque de Holanda. New Aurélio dictionary. 2. ed. Rio de Janeiro: Nova Fronteira, 1993.

OLIVEIRA, RAMON. Informática educativa: dos planos e discursos à sala de aula. Campinas, SP: Papirus, 1998.

PAIS, L.C. Didática da Matemática: uma análise da influência francesa. Belo Horizonte: Autêntica, 2002.

PANIZZA, Mabel (Org). Teaching maths in early childhood education: analysis and proposals. Porto Alegre: Artmed, 2006.

PAPERT, S. A máquina das crianças: repensando a escola na era da informática. Porto Alegre: Artes Médicas, 1994.

PAVANELLO, M. R. The abandonment of the teaching of Geometry: a historical view. 1989. Dissertation (Master's in Education) State University of Campinas, Campinas, 1989.

PEREIRA, Maria Regina de Oliveira. School geometry: an analysis of studies on the abandonment of its teaching. 2001. Dissertation (Master's in Education.) Pontifical Catholic University of São Paulo, São Paulo, 2001.

PINTO, N. B. Didactic contract or Pedagogical contract? Revista Diálogo Educacional, Curitiba, v. 4, n. 10, p. 93-106, Sep./Dec. 2003.

RODRIGUES, M. H. W. L.; BRAVIANO, G. Dynamic geometry: a new geometry? Revista do Professor de Matemática, São Paulo, n. 49, May/August, p.22-26, 2002.

SANCHO, Juana M.; HERNANDEZ, Fernando. Technologies to transform education. Education.Porto Alegre:Artmed,2006.

SILVA, Maria Célia Leme da. Thales' theorem: a didactic engineering using Cabri-Géomètre II.

1997 Dissertation (Master's in Maths Teaching) Pontifical Catholic University of Paraná, São Paulo, 1997.

SILVA, Sônia Firette Nunes da. Geometry in the early grades: why not? The choice of content - a task that reveals teachers' ability to decide. 2006. Dissertation (Master's in Education) Federal University of Paraná, Curitiba, 2006.

SILVA, E. O; MOREIRA, M.; GRANDO, N. I. O contrato didático e o currículo oculto: um duplo olhar sobre o fazer pedagógico.Revista Zetetiké, Campinas, SP, n. 6, p. 9-23, jul/dez, 1996.

USISKIN, Z. Solving the permanent dilemmas of school geometry. In: LINDQIST, M. M.; SHULTE, A. P. (Org.). Aprendendo e ensinando geometria. São Paulo: Atual, 1998.

VALENTE, J. A. V. (org). Computers and knowledge: rethinking education. 2. ed. Campinas, SP: Unicamp/Nied, 1998.
_____ . The computer in the knowledge society. Campinas, SP: Unicamp/Nied, 1999.

_____ . History of the arrival of computers in Brazil. Available at: <http://www.puc-rio.br/sobrepuc/depto/psicologia/download/pdf/Ana%20Maria%20Nicolaci- da-Costa.pdf> Accessed on: 12 May 2007.

_____ . Informatics in education in Brazil: Analysis and historical contextualisation. Available at: <http://maresias.ufla.br/~remulo/com202/arquivos/cap1.pdf>. Accessed on: 04 October 2007.

_____ . Research, communication and learning with the computer: The role of the computer in the teaching-learning process. In: ALMEIDA, Maria Elizabeth Bianconcini de; MORAN, José Manuel (Org.). Integration of technologies in education. Brasília: Ministry of Education, 2005. p. 22-31.

_____ .; ALMEIDA, Fernando J. de. Analytical view of informatics in education in Brazil: the question of teacher training. Revista Brasileira de informática na educação, n. 1, 1999. Available at: <http://www.inf.ufsc.br/sbc-ie/revista/nr1/valente.htm>. Accessed on: 06 Dec. 2007.

ZUIN, Elenice de Souza Lodron. National Maths Curriculum Parameters for the 3rd and 4th Cycles of Primary Education and the teaching of Geometric Constructions, among other considerations. In: ANPED ANNUAL MEETING, 25, 2002, Caxambu, MG. Annals... Available at: < http://189.1.169.50/reunioes/25/excedentes25/elenicezuint19.rtf>. Accessed on: 08 October 2007.

_____ . The ruler and the compass: geometric constructions as school knowledge in Brazil. 2001. 206 f. Dissertation (Master's in Education) Faculty of Education, Federal University of Minas Gerais, Belo Horizonte, 2001.

ZULATTO, R. B. A. Maths teachers who use Dynamic Geometry software: their characteristics and perspectives. Dissertation (Master's in Education) Universidade Estadual Paulista Julio de Mesquita Filho, Rio Claro, 2002.

Printed by Books on Demand GmbH, Norderstedt / Germany